GUSTATION AND OLFACTION

FOOD SCIENCE AND TECHNOLOGY

A SERIES OF MONOGRAPHS

Editorial Board

Maynard A. Amerine, Rose Marie Pangborn, and Edward B. Roessler, PRINCIPLES OF SENSORY EVALUATION OF FOOD. 1965.

C. R. Stumbo, THERMOBACTERIOLOGY IN FOOD PROCESSING. 1965.

Gerald Reed, ENZYMES IN FOOD PROCESSING. 1966.

S. M. Herschdoerfer, QUALITY CONTROL IN THE FOOD INDUSTRY. Volume I—1967. Volume II—1968. Volume III—in preparation.

Hans Riemann, FOOD-BORNE INFECTIONS AND INTOXICATIONS. 1969.

Irvin E. Liener, TOXIC CONSTITUENTS OF PLANT FOODSTUFFS. 1969.

Leo A. Goldblatt, AFLATOXIN: SCIENTIFIC BACKGROUND, CONTROL, AND IMPLICATIONS. 1969.

Martin Glicksman, GUM TECHNOLOGY IN THE FOOD INDUSTRY. 1969.

Maynard A. Joslyn, METHODS IN FOOD ANALYSIS, second edition, 1970.

A. C. Hulme, THE BIOCHEMISTRY OF FRUITS AND THEIR PRODUCTS. Volume 1—1970. Volume 2—in preparation.

G. Ohloff and A. F. Thomas, GUSTATION AND OLFACTION. 1971.

GUSTATION AND OLFACTION

An International Symposium
Geneva, June 1970

Sponsored by
Firmenich et Cie

Edited by

G. OHLOFF AND A. F. THOMAS

Firmenich et Cie, Geneva, Switzerland

1971

ACADEMIC PRESS · LONDON · NEW YORK

ACADEMIC PRESS INC. (LONDON) LTD
Berkeley Square House,
Berkeley Square,
London, W1X 6BA

U.S. Edition published by
ACADEMIC PRESS INC.
111 Fifth Avenue,
New York, New York 10003

Library of Congress Catalog Card Number: 78–153530
ISBN: 0–12–524950–0

PRINTED BY OFFSET IN GREAT BRITAIN BY
WILLIAM CLOWES AND SONS LIMITED, LONDON,
BECCLES AND COLCHESTER

ACKNOWLEDGMENTS

Firmenich & Cie wish to express their gratitude to everyone present at the Symposium on Gustation and Olfaction, more particularly to :

Prof. Dr. L. Ruzicka – Honorary Professor at the Swiss Federal Institute of Technology, Zurich
Chairman of honor of the Symposium

Prof. O. Reverdin – Chairman of the Swiss Council of Scientific Research
Chairman of the Symposium

Prof. M.R. Kare – Philadelphia, USA
Vice-Chairman of the Symposium

Prof. J. Le Magnen – Paris, France
Vice-Chairman of the Symposium

Prof. D. Ottoson – Stockholm, Sweden
Vice-Chairman of the Symposium

They also wish to extend thanks to their Geneva collaborators and members of the organizing committee : Dr. G. Ohloff, general secretary, Dr. E. Sundt, assistant general secretary, Mrs. O. Schaub and Mr. A. Delor, as well as to Miss Y. Kane from Firmenich Inc., New York, for their valuable contribution to the success of the Symposium.

FOREWORD

An industrial enterprise conscious of its responsibilities cannot but promote scientific advance in its field of activities by every possible means. Failing such an obligation, an enterprise would not enjoy the freedom of full exploitation of the results of research. This obligation was acknowledged by Dr. Roger Firmenich when, in his address of welcome, he expressed his faith in science. Following a scientific tradition of more than half a century Firmenich chose to celebrate its 75th anniversary by organizing and sponsoring an International Symposium on Gustation and Olfaction. Just as important a celebration was the 50 year old friendship and successful collaboration between Firmenich and the pioneer of modern chemistry of natural products, Professor Leopold Ruzicka. He was, in fact, the first research director (1925) of the enterprise (then called Naef & Cie) before he went to Utrecht and thereafter to Zurich where, for a long time, he was the leading spirit inspiring research at the chemical laboratory of the Swiss Federal Institute of Technology. In his "Outline of the History of Firmenich Research", Dr. G. Ohloff, now research director of Firmenich, acknowledged the merits of this great scientist in the following terms : "Bien que son séjour à Genève n'ait été que de courte durée, le professeur Ruzicka a insufflé à la recherche un esprit qui fut cultivé et développé par Max Stoll, son élève et son successeur. Cet esprit est resté très vivant depuis bientôt 50 ans et nous souhaitons ardemment qu'il ne meure jamais."

In his introduction Professor Le Magnen pointed out that 25 years ago a symposium of this type would not have been thought of, not only because the present knowledge in the field of electrophysiology was not available, but also because only minor importance for human life was attributed to both chemical senses in general and to the sense of smelling in particular.

As shown by the increasing number of meetings of scientists who investigate the fundamental problems of chemoreception, this field of interdisciplinary science is on the verge of a revolutionary phase. However, as was to be expected, this symposium, like others before, did not provide solutions to the problems discussed. It is not even certain whether we have got any nearer to the target. More weight was nevertheless given to facts than to theories. In the future hypotheses should be more carefully supported by conclusive results. The symposium showed encouraging signs of this tendency.

Geneva, November 1970

G. Ohloff

CONTENTS

DETECTION AND RECOGNITION OF ODOR MOLECULES

David G. Moulton

Monell Chemical Senses Center and Department of Physiology,
University of Pennsylvania, and
Veterans Administration Hospital, Philadelphia, Pennsylvania 19104.

INTRODUCTION

The molecular bases of our ability to detect and recognize odors are the central problems in olfaction. We have no generally accepted descriptions of these processes. Indeed, we lack a full understanding of even the most elementary aspects of receptor function. This is not to say that overall progress in this field has been slow. On the contrary, we have – in recent years – learned much about the ultrastructure of odor receptors, the electrical properties of the olfactory epithelium, and the role of pheromones (and other compounds with signalling properties) in controlling behavior and reproductive processes. Also, we have elegant analyses of the circuitry of the olfactory bulb and the properties of odor receptors in certain species of insects (to take but a few examples).

Where there has been limited growth is in our understanding of the odorant-receptor interaction. The reasons for this are several and – by way of an introduction – worth brief consideration.

The most serious constraint is that existing techniques are often inadequate. For example, it is difficult to measure electrical events across the membrane of a receptor whose largest diameter is only 5-7 microns or so. Sustained intracellular recordings are rare and have yet to yield much information. Fortunately, receptor cells with ten times the area of those previously studied exist in at least one amphibian (*Necturus*, the mud puppy) but until they are exploited, many questions concerning receptor function and electrogenesis will remain unresolved. Single fiber analysis – a technique first applied to the frog sterno-cutaneous muscle by Adrian and Zotterman in 1926 – offers a further powerful approach to the quantitative definition of receptor response. Unfortunately, the primary olfactory neurons – with a mean diameter of only 0.2 microns – are among the smallest axons of the body. Indeed, they are individually beyond the resolving power of the light microscope. Consequently the physical isolation of single fibers is not feasible with existing techniques.

Analysis is further confounded by inability to isolate receptors physically from other cell types. Unlike the retina, where nerve cells predominate (and even all-cone or all-rod regions can be found) the overwhelming bulk of the olfactory epithelium consists of non-neural elements – the basal and supporting cells. This complicates interpretation of biochemical and other analyses. Surprisingly, however, there have been very few attempts to exploit such modern biochemical and histochemical approaches as are available. Nor has there been systematic use of the techniques and concepts of such disciplines as molecular pharmacology and immunochemistry in investigations of the odorant-receptor interaction. While this situation may be changing there is no assurance that progress will be rapid. After all, there have been over many years extensive investi-

gations of the structure of plasma, mitochondrial and other membrane systems. But there is
still little agreement about the detailed architecture of any membrane, let alone one as complex
as that of the olfactory receptor.

The inaccessibility of the organ also creates a technical obstacle. We deliver odorants
to the nose at a known concentration and flow rate, then measure neural or behavioral response.
But a proportion of the molecules may never reach the receptors. They are either adsorbed to
the walls of the nasal chamber, bypass the olfactory cleft or fail to hit the receptors (see Stuiver,
1958). The precise number of molecules that ultimately contact the olfactory surface thus depends, in part, on the aerodynamical properties of the nasal chamber. But these properties are
poorly understood. The ebb and flow of air movement induced by normal inspiration and sniffing results in a complex distribution of odorous molecules across the elaborate convolutions of
the olfactory sheet. This action is not easily studied in model systems. Nor do models reproduce
continuing and cyclical alternatives in the degree of engorgement of the nasal mucosa – changes
which may significantly alter flow patterns in the nasal airways. Another aspect of accessibility
is the chemical alteration of the odorant molecule by the action of the nasal mucus. Nicollini
(1954), believed that the nasal cavity is a veritable chemical laboratory. He even argued that some
forms of anosmia resulted from the chemical action of the mucus. While this extreme view may
be questioned, the chemical properties of the secretions covering the olfactory surface may, nevertheless, influence receptor actions. This sheet appears to consist of both aqueous and viscid
layers and contains oxidative and hydrolytic enzymes, mucopolysaccharides, and pigment granules (Reese, 1965; Shantha and Nakajima, 1970; see also Moulton and Beidler, 1967).

Even the extreme sensitivity of the nose itself imposes limitations. There is a report
that a trained odor panel detected contaminants in menthol, citral, methyl salicylate and safrol
at concentrations several powers of ten below those detected by the flame ionization detector
of a gas chromatograph (Kendall and Neilson, 1964). While it is true that the relative sensitivities of physical and biological detectors are in some cases a matter of controversy, one point is
clear : standards of purity acceptable to the organic chemist may not be acceptable to the olfactory organ. As an unfortunate result much of the literature concerning structure-activity relations is, at best, of doubtful value. It can be argued that improvements in instrumentation and
the introduction of new approaches, such as plasma chromatography, hold promise of reducing
this gap. But even gas chromatography is routinely used by no more than a handful of those investigating olfactory response.

We also face a more general problem : knowledge, or its significance, is not always
evident to those who could best use it. At the turn of the century, the only instrument available
for recording neural response was the capillary electrometer, and the only nerves whose response
was slow enough to record with this instrument were the primary olfactory neurons (Garten,
1900). But those who studied this response were not primarily interested in olfaction, and those
interested in olfaction may not have known of the instrument. In fact, it was more than half a
century before the electrical activity of primary olfactory neurons was recorded in a continuing
investigation of olfactory mechanisms. A further example is the report by Gerebtzoff in 1953
that certain elasmobranchs have unusually large receptors. Electrophysiologists have yet to exploit this observation. Similarly, Guillot, in 1948, published accounts of his investigations on partial or specific anosmia. However, it was not until 1967 that Amoore, also realizing the potential
of this condition as a key to understanding certain structure-activity relations began more systematic studies on "odor blindness".

There is one final difficulty : too often a concept becomes accepted as established, or highly probable, before it has met adequate criteria of proof. Further research may then seem less important or is misdirected. To varying degrees this seems to have happened at one or more points in the development of ideas concerning the location of active sites on the olfactory receptor, the role of pigment in olfaction, the renewal of cell populations in the olfactory epithelium, and the role of receptor specificity in determining differential sensitivity to odors. These are all areas where initial findings have led some to favor a certain interpretation. Later evidence has cast doubt on the earlier assumptions and the problems have generally turned out to be more complex and less susceptible to solution than first appeared.

In what follows we shall explore these examples further. Not only do they offer a window onto the uncertain nature of much of our knowledge in this field, but they will also serve as a framework for outlining some recent evidence and defining meaningful questions (an essential pre-requisite for fruitful research).

At the outset however, may I remind you of the structure of the olfactory epithelium, since much of what follows assumes knowledge of its basic features (Fig. 1).

LOCATION OF RECEPTOR SITES

It is a plausible assumption that molecules are first weakly and reversibly adsorbed to receptor sites on, or near, the surface of the cell, as suggested in essence by Stoll (1965) and others. But where, precisely, are these sites? Are they evenly distributed over the surface or restricted to one of a few specific regions? Do they form a scattered mosaic of membrane parts ("generator units") functioning independently of each other, as has been proposed for another sensory receptor – the unmyelinated nerve terminal of the pacinian corpuscle (Lowenstein and Rathcamp, 1958)? If so, do the various parts have different properties or are they interchangeable? Do such sites contain a specific type of receptor molecule that is involved in binding the odorant? To the bio-chemist concerned with isolating a specific subcellular fraction on the assumption that it contains such receptor molecules, these are questions of immediate practical significance. There would be little point, for example, in isolating ciliary membranes for analysis, if they should prove to have no essential function in the primary transduct on process. What evidence is available that might clarify these points?

If we view the surface of the olfactory epithelium through the scanning electron microscope we are immediately struck by the dense array of filamentous projections (Fig. 2).These are cilia – extensions of the membranes of the receptor cells – and in life they lie in an aqueous sheet of pigmented secretions. They present an immense surface area vastly exceeding the 3 sq. cm. or so that is given as the area of the olfactory epithelium in man (Dieulafé, 1906). Since they are presumably the first structures encountered by odor molecules, they might seem the most logical surfaces on which to find active sites.

Consistent with this view is the unique form and behavior of olfactory cilia. Typical motile cilia beat in coordinated rhythmic waves. Such cilia occur on the respiratory epithelium which borders the olfactory epithelium. Their rapid movements propel the mucus sheet at a rate of about 10 mm/min both in chickens and man (Bang and Bang, 1963; Hilding, 1932 a; 1932 b). Even in sections prepared for light or electron microscopy the ordered alignment of respiratory cilia necessary to coordinate the propulsive action is often partially preserved (Fig. 3). In contrast, any movement so far seen in olfactory cilia is slow, disorganized and ineffective in transporting mucus. In fact it may be an abnormal response (see Reese, 1965).

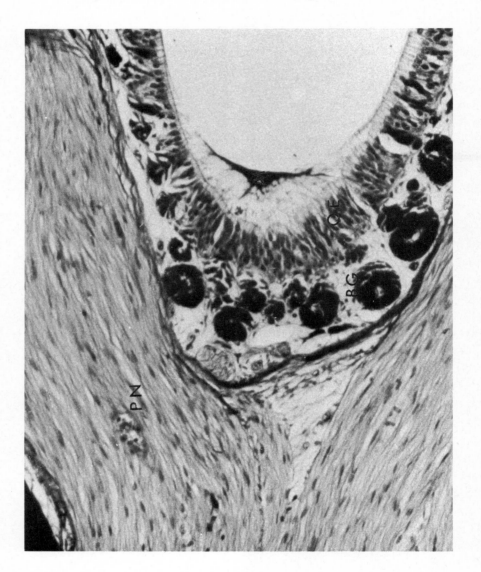

Fig. 1. Structure of the olfactory epithelium.

a) Relation of epithelium to surrounding structures. This Saggital section of a frog's head shows the massive bundles of primary neurones (PN) converging towards the olfactory bulb and the unevenly distributed Bowman's glands (B.G.) whose secretions – remnants of which can be seen (M) – bathe the surface of the olfactory epithelium (O.E.) (Mallory stain 160X.)

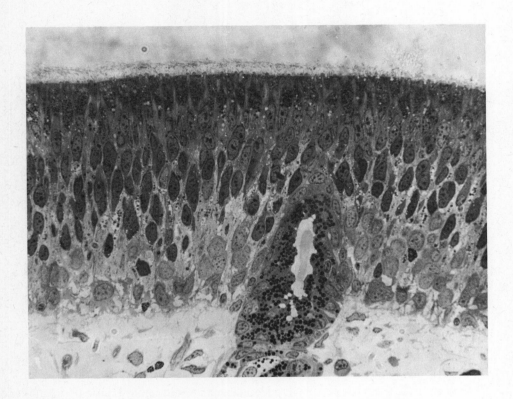

Fig. 1. b) Section of frog's olfactory epithelium.
(Courtesy P.C.C. Graziadei).

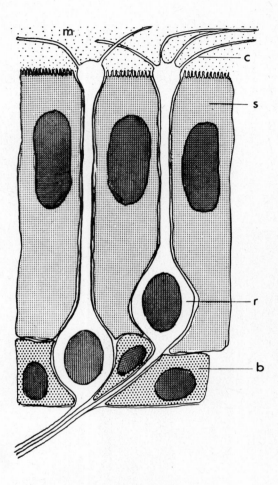

Fig. 1. c) Simplified schematic diagram to illustrate the main features of the olfactory epithe-
     lium (compare with Fig. 1. b).

The three main cell types commonly identifiable are the receptor (r), supporting (s) and
basal (b) cells. The receptors bear cilia (c). Some workers have described a fourth cell
type in the dog and in the African white-tailed rat (Okano et al., 1967; Kauer, 1969),
but it is not shown in this figure. The receptors are bipolar neurons. One pole, the dendrite or transducing element, is in contact with the mucus (m) sheet. The other is an
axon or conducting element, which transmits nerve impulses to the olfactory bulb. The
receptors are embedded among supporting cells which have several times the volume of
the receptors. They are thought to have a secretory function in some species and are
intimately related structurally and functionally with the receptors. (In other words, they
are probably far from having the merely supportive role that their name suggests). The
nuclei of the supporting cells are found in the outermost region of the epithelium while
those of the receptor cells lie immediately below. Finally, the basal cells form the deepest
layer and rest on the lamina propria. Within this lamina are found the Bowman's glands
whose ducts reach the olfactory surface onto which they inject their pigmented secretions.

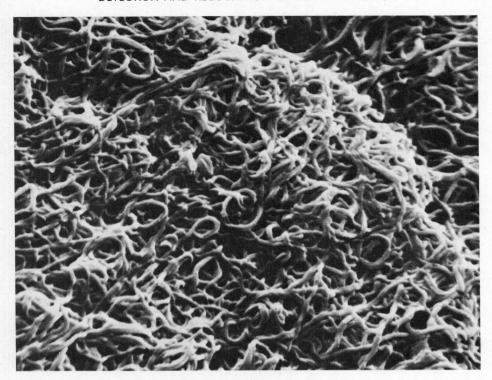

Fig. 2. Surface of olfactory epithelium of frog showing an apparently random distribution of cilia 7500X. (Courtesy P.P.C. Graziadei).

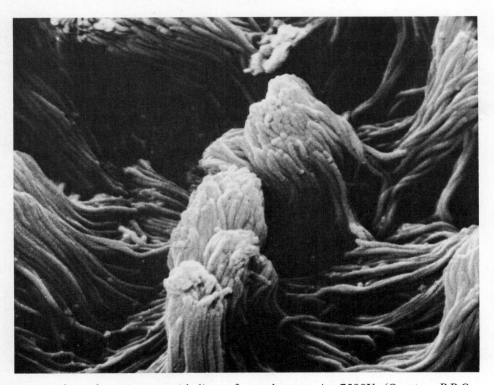

Fig. 3. Surface of respiratory epithelium of a gopher tortoise 7500X. (Courtesy P.P.C. Graziadei).

In many vertebrate species the ciliary length exceeds 10 microns and may be 80 microns or more. This enables the cilia to penetrate far into the mucus sheet. Indeed in some cases (such as in certain species of frogs in which lengths of up to 200 microns have been reported) the upper segments of the cilia may lie along the interface between the viscid and aqueous layers of the mucus sheet thus enhancing exposure to odor molecules (see Reese, 1963; Moulton and Beidler, 1967; Steinbrecht, 1969). As seen in Fig. 4 – a silver-stained section of frog epithelium – the cilia (or features associated with them) fan out from the terminal knob of the receptor and appear to span almost the entire width of the secreted sheet. It should be remembered, however, that in life this sheet may be thicker.

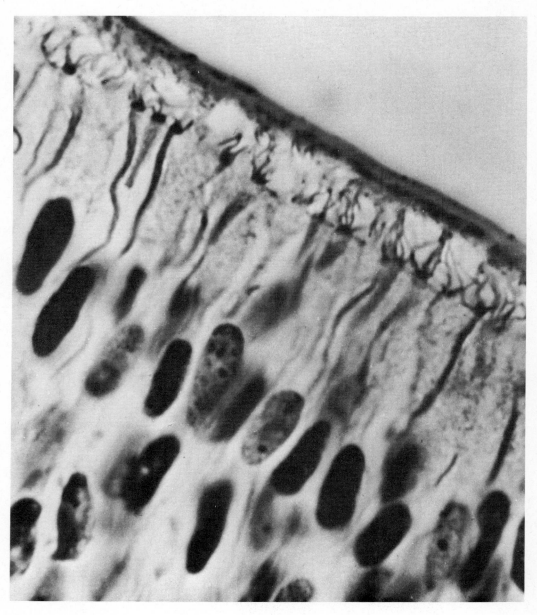

Fig. 4. Olfactory epithelium of frog to show terminal knobs bearing stumps of cilia which project into the mucus sheet. Bodian stain 2300X.

To gain a more detailed view of the receptor and its relation to other epithelial elements, consider an electron micrograph. Fig. 5 shows the peripheral region of the olfactory epithelium of a rabbit. The receptors terminate in knobs bearing a few cilia. Because cilia splay out from the knob at various angles they pass out of the plane of the section. Consequently they are generally sectioned close to the knobs and appear as short stumps. Their full length is thus seldom, if ever, represented. The thin sections required for electron microscopy also make it difficult to estimate the total number of cilia. Over 20 per cell is probably not an uncommon number for mammals, and over 100 per cell has been reported for the dog (Okano et al., 1967). The microvilli on the supporting cells are also visible in the figure. Since they are thinner than cilia and tend to lie vertical to the surface, they are probably seen at their full length. Microvilli lack the internal structure of cilia (a pair of central tubules surrounded by a circle of nine pairs of peripheral tubules).

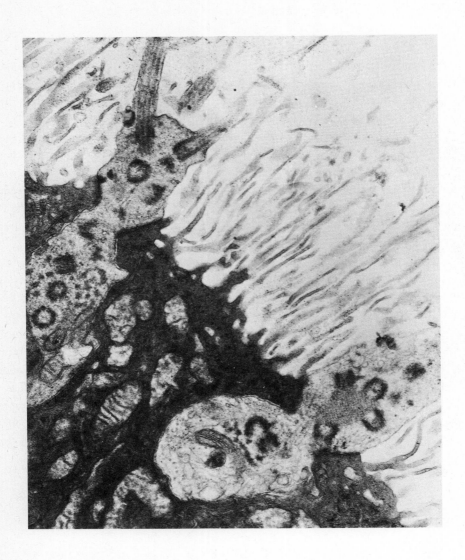

Fig. 5. Olfactory epithelium of rabbit 32000X.

The degree of interrelation between neighboring elements in this system is not yet clearly established. Interest centers on regions where contiguous cells come into close apposition – the so-called tight junctions (*zonulae occludentes*). Such junctions (when they are not considered to be artifacts) are thought to have one of several possible functions depending on their precise structure and location (Brightman and Reese, 1969). Examples of these complexes appear in Fig. 5. One of their functions may be to bar the movement of molecules along intercellular spaces. If the junctions between the apical segments of cells in the olfactory epithelium were to act in this way, they could be important in limiting access of odor molecules to the receptor surface. A study by Reese and Brightman (1970) on monkeys suggests that this may indeed happen. They injected horse-radish peroxidase into the circulation and applied appropriate histochemical techniques to detect the reaction product. It was found that the product does not reach the olfactory surface. Its passage is effectively blocked by continuous belts of tight junctions encircling the receptor and supporting cells and sealing off the underlying cell interspaces from the overlying secretions. This observation might be taken as further evidence of the importance of the ciliated region of the receptor in the primary transduction process.

The evidence, then, would seem to bear out the idea that the location and dimensions of cilia favor maximum contact with incoming molecules and that they may therefore carry the essential receptor sites. However, there are certain additional findings that foster caution. Since they relate, in part, to the vomeronasal organ we will consider its properties briefly.

The vomeronasal organ is present in most land-dwelling vertebrates although rarely functional in man. It lies in a blind-ending sac which opens onto the floor of the nasal chamber in many mammals. In so far as the evidence takes us, its range of responsiveness to odors, and its ultrastructure, appear comparable to that of the olfactory organ, although there are some exceptions (Tucker, 1963; Altner and Muller, 1968; Bannister, 1968; Graziadei and Tucker, 1968; Kauer and Moulton, 1970). One of these exceptions can be seen in Fig. 6, which is an electron micrograph of the vomeronasal epithelium of the rabbit.

Each receptor has a terminal knob and there are junctional complexes similar to those in the olfactory epithelium. However, instead of cilia the receptors bear large numbers of microvilli (Kauer and Moulton, 1970). Microvilli (or villous tufts) instead of cilia have also been found on one type of receptor of the minnow *Phoxinus phoxinus* (Bannister, 1965), in the guitar fish, *Rhinobatus lentiginous* (Reese and Brightman, 1970), as well as on the vomeronasal receptors of other species. In addition, microvilli in combination with cilia occur on the olfactory receptors of some species of birds, and in the fish *Crassius crassius* (Andres, 1968; Brown and Beidler, 1966; Graziadei and Bannister, 1967; Wilson and Westerman, 1967). This evidence suggests that while some extension of the membrane may still be important in nasal chemoreception, cilia, as such, are not necessarily essential.

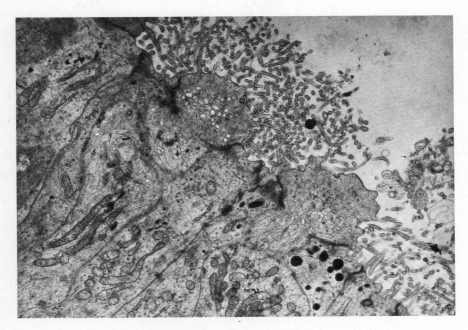

Fig. 6. Vomeronasal epithelium of rabbit 32000X.

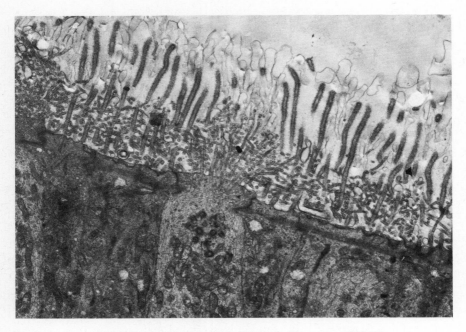

Fig. 7. Vomeronasal epithelium of albino rat 32000X.

A further example of vomeronasal epithelium is seen in Fig. 7. The unusual feature here is that the receptor microvilli are not only thinner but also shorter than those on the contiguous supporting cells (Kauer and Moulton, 1970). If the criteria of greatest accessibility to odorous molecules were applied to determine the location of active receptor sites, the supporting cell would have to be considered as the true receptor. However, it is difficult to accept the idea that the supporting cell may act as an accessory cell (comparable to a taste cell) and transmit excitation to the neuron through specialized regions of contiguous membranes. Nevertheless, it is a curious comment on the status of existing knowledge that despite its seeming improbability, such a basic challenge is difficult to refute.

It might be argued, with some justification, that it is of little consequence whether the membrane is extended into projections called cilia or microvilli, providing a major increase in surface area is achieved. The findings of Tucker, (1967), however, place doubt on even this assumption. He found some indication (in a preliminary report) that even removal of the majority of olfactory cilia (from receptors lacking microvilli) does not significantly alter the response to odor stimulation recorded from primary neurons (Tucker, 1967).

In summary, cilia are not as essential to olfaction as once supposed. In particular, it is not clear that their internal structure has any unique significance in the primary transduction mechanism. Active sites can occur on surfaces of the terminal knob free of cilia, and, if excited, are probably capable of firing a nerve impulse. At apical poles of receptors, encircling belts of tight junctions may restrict access of odor molecules to the terminal knob and immediately adjacent membrane, although further investigations of the origin and distribution of these structures are needed.

Nevertheless, it would not be surprising if any region of the receptor membrane accessible to odorants bears receptor sites. Projections such as microvilli and cilia may serve to increase the efficiency of the system by providing a larger surface area, placing it nearer the overlying airway, and thus enhancing the probability of contact with odorous molecules.

Knowledge of the location of receptor sites is one step in understanding the olfactory transduction process. A further step is to determine the nature of the receptor substance. Unfortunately studies directed towards this goal are few. Historically, at least, the starting point has been the olfactory pigment.

THE OLFACTORY PIGMENT COMPLEX

The dearth of evidence on the biochemistry of the olfactory epithelium is not entirely surprising when we remember that, with the outstanding exception of the retina, our understanding of the biochemistry of any receptor organ is not well advanced. But in the case of the retina, an early and potent focus of research interest was the obvious presence of color in the organ. Later work by Wald and others has established that carotenoid – protein pigments are essential to vision.

Noting the yellowish or brownish color of the olfactory epithelium in many vertebrates,several workers have argued that pigment may also play a role in olfaction. Indeed, it is over a century ago that Ogle (1870) first cautiously suggested that pigment, if not essential to olfaction, "at least conduces much of its keenness and perfection". More recently, support for this view has come from several sources. For example, it is sometimes held that albino animals are anosmic since they are unable to detect poisonous plants by smell. But if we trace this belief to its origin in Darwin's (1875) treatise on "The Domestication of Plants and Animals" we find a misinterpretation. Darwin merely noted that certain vegetable poisons, when eaten, cause ill effects in unpigmented animals but not in pigmented members of the same species. He says nothing about smell. In fact, he was almost certainly citing examples of photodynamic sensitization.

Far from showing anosmia, albino rats are at least as sensitive to the odors of members of an homologous series of aliphatic alcohols as are their pigmented litter mates. Furthermore no differences could be detected in either the color or diversity of pigment fractions derived from a chromatographic separation of albino and pigmented rat olfactory epithelia (Moulton, 1962). Of course these arguments do not eliminate the possibility that pigment is involved in olfaction. They merely demonstrate that the assumption is not as firmly based as once supposed.

In fact since Ogle's speculations, there have been several theories implicating pigment in the olfactory transduction process, including one by Wright et al. (1956). The most recent of these is that of Rosenberg et al. (1968). It takes as a starting point the observation that the olfactory epithelium of cattle contain large concentrations of beta-carotene (Milas et al., 1939; Briggs and Duncan, 1961; Moulton, 1962; Kurihara, 1968). Now, it is known that all-*trans* β-carotene powder has semiconductor properties. When various odorous gases were passed over the powder in a conductivity cell, the semiconductor current increased by factors up to 10^7 (Rosenberg et al. 1968). The current increase depended on the amount and nature of the gas and appeared to be the result of a weak bound donor-acceptor complex (Misra et al., 1968). Rosenberg and his associates also reported finding a correlation with odor intensity and current increases due to adsorption of the same gases on the β-carotene powder. On the basis of these findings, the authors proposed that pigments in the olfactory receptor cell membrane constitute the receptor sites.

Despite some attractive features, this theory faces certain difficulties. When odorous gases were adsorbed to the surface of the β-carotene crystal the time for current to rise to maximum amplitude was in the order of minutes. The corresponding neural response, however, would be expected to reach maximum amplitude in milliseconds. Furthermore oxygen and carbon dioxide elicited marked current increases although they have no odor for man.

More critical, however, is whether β-carotene is widely distributed throughout representative vertebrate species. or restricted to certain groups as some evidence suggests (Moulton, 1962). In fact, even the common assumption that the yellow of the olfactory epithelium is due to carotenoid pigments is incorrect. Kurihara (1967), using chromatographic techniques, isolated four chromoproteins from bovine olfactory epithelia. He estimated that the ratio of the contributions of chromoprotein in pigment granules, water soluble pigments, and pigments soluble in inorganic solvents, to the color of the olfactory epithelium is 100:15:1·2. Thus any theory implicating pigment in the olfactory mechanism must take into account the fact that the bulk of the pigment, if not all, is composed of non-carotenoid compounds. It must also contend with failure of most histologists to find pigment in the receptor cell. Instead, the pigment appears to be concentrated in the supporting cells and Bowman's glands. Finally, the apparent absence of pigment from some species of fish, and the known frequency with which pigmented compounds are found throughout the body (often in structures such as ear wax, where they appear to meet no obvious functional requirement) throws some doubt on the idea that pigments have any important significance for olfaction (see Moulton, 1970).

On the other hand, non-carotenoid chromoproteins are known to be involved in a number of physiological processes. It is therefore not entirely inconceivable that small concentrations of such compounds act as receptor substances. The weak binding of an odorant to these sites might be sufficient to induce a change in the shape of the non-protein part of the chromoprotein molecule. The resulting conformational change in the protein moiety could lead to a transient alteration in the permeability of the receptor membrane allowing for an interchange of ions across it. When the resulting redistribution of charge across the membrane achieves a critical magnitude, it could initiate the events that lead to the generation of a nerve impulse at or near the receptor axon hillock (Moulton, 1970).

In summary, it appears that the olfactory epithelium in certain species of mammals contains retinol, beta-carotene and at least four non-carotenoid pigments claimed to be chromoproteins. The yellowish-brown color of the olfactory epithelium in most vertebrates seems to be due to the non-carotenoid pigments. It is not clear whether the pigment is present in the receptor cell and the bulk, at least, is concentrated in the supporting cells and the cells of Bowman's glands. Some species of fish appear to lack the pigment. These are difficulties in accepting existing theories implicating pigment in the primary olfactory transduction process. However, the dense concentrations of olfactory pigment in some species is curious, and the possibility that pigment may play some role in olfaction remains.

PROLIFERATION KINETICS OF CELL POPULATIONS IN THE OLFACTORY EPITHELIUM

When cells of our brain are destroyed they cannot be replaced. Similarly, in the retina, we are born and die with the same rods and cones. And in the inner ear, the Organ of Corti has no power to replace damaged hair cells. However, some tissues of the body, such as those facing the impact of continuous mechanical traffic, behave differently. The cells of the taste bud, for example, undergo continuous renewal. In the rat, epithelial cells surrounding a bud continuously divide and enter the bud at a rate of about one every 10 hours. The average life of the taste cell is about 11 days (Beidler and Smallman, 1965; Conger and Wells, 1969). This turn-over raises some interesting questions. For example, how does the replacement cell acquire properties that preserve our capacity to make taste discriminations?

But the taste cell is an accessory cell, not a nerve cell. Its death leaves intact the fiber linking it to the brain. The fiber is then free to innervate a replacement cell. Not so the olfactory receptor; it is a biopolar neuron in which no synapse separates the transducing and impulse conducting elements of the cell. Should it die, any replacement cell would have to establish its own axonal connection with the olfactory bulb. To do so, the axon must grow caudally (up to several centimeters) before reaching what may turn out to be a rather specific site in the bulb.

Such "homing" of the sensory axons would not be exceptional since it is shown, for example, by regenerating axons in frog optic nerve. However, we are concerned here not with regeneration – a response to an experimentally or pathologically induced destruction of the axon – but with continuous and normally occurring death and renewal of entire neurones in the adult. The idea that this can occur goes against classical assumptions. In this context the continuous renewal of olfactory receptors would seem unlikely and the olfactory epithelium might be expected to behave more like the retina than the taste bud.

However, there have long been suggestions that the basal cells are continually dividing and that the products of cell division replace supporting cells. Recently, Shantha and Nakajima (1970) have argued against this view. In a histochemical study of the olfactory mucosa of the

Rhesus monkey *(Macaca mulatta)* they have found that the upper part of the dendrite process and terminal knobs and the supporting cell show high oxidative enzyme activity. The basal cells, however, show high hydrolytic enzyme activity and low oxidative enzyme activity. Furthermore, the basal cells show very strong adenosine triphosphatase activity while the supporting cells and receptors show much less. The authors conclude that basal cells cannot give rise to supporting cells and are not young forms of supporting cells.

Andres (1965, 1966, 1968) has taken a different view. As a result of electron microscope studies on the olfactory epithelium of cats, dogs and rats, he concluded that the entire epithelium is continually being renewed. More recently, however, (Andres, 1970) he has withdrawn this suggestion in so far as he finds mitosis to be rare among receptor cells in the adult.

To gain more decisive evidence on this point and, in particular, quantitative information on the behavior of the relevant cell populations, it is necessary to be able to identify dividing cells and deduce their movements. This can be done by using tritiated thymidine, a radioactive precursor of desoxyribonucleic acid (DNA) which is actively taken up by cells synthesizing DNA prior to cell division. Cells labelled by this tracer can be identified by autoradiographic techniques and light microscopy.

We have recently exploited this approach to derive a preliminary definition of the dynamic structure of these cell populations in mice olfactory epithelia (Moulton et al., 1970). The mice were injected with tritiated thymidine and different groups were killed at various periods later (20 mins., 1 hr., 2 hrs., 1, 4, 8, 16, 20, 30, 44 and 90 days).

It is immediately apparent, after 20 minutes survival, that a significant population of cells undergo division (Fig. 8). About 90 per cent of these labelled cells are scattered throughout the basal cell layer. The remainder lie more peripherally in the zones occupied by the nuclei of the supporting and receptor cells. From this we can deduce two points. First, either supporting or receptor cells, or both, undergo cell division *in situ.* However, they do so at a rather slow rate, typical of a population that is actively expanding proportionately to the rate at which body weight increases. This would not be surprising in rodents.

The second point is that the number of cells in the basal zone is more characteristic of a proliferating than of an expanding cell population. In other words, the rate of division exceeds that required by a growing tissue and implies that cells are dying and being replaced. If this deduction is correct we might expect to find that some of the products of cell division would migrate to occupy new positions as is characteristic of proliferating cell populations. In fact, the proportion of labelled nuclei in the basal cell zone declined to 23 per cent between the 4th and 20th day of survival (Fig. 9). There is a corresponding increase in the cells lying more peripherally. In other words, labelled cells migrate outward at a rate consistent with continuous cell renewal and replacement (Fig. 9).

The central question is whether the cells that are undergoing renewal are receptor, or supporting cells, or both. Although conclusive evidence is lacking, there is some indication that receptor cells are being replaced. When migrating products of basal cell division come to rest their nuclei lie primarily in the mid zone of the epithelium. This is the region occupied by the nuclei of the receptor cells. There is no evidence that supporting cell nuclei are ever found in this region, and certainly not with the frequency that labelled nuclei are seen.

There is also other evidence that adds some support to the view that the receptor cells undergo continuous renewal. For example, Thornhill (1967) observed in the lamprey *(Lampetra fluviatilis)* occasional anons darker than surrounding ones in regions of the olfactory epithelium. This suggests the presence of degenerating fibers. Furthermore Shantha and Nakajima

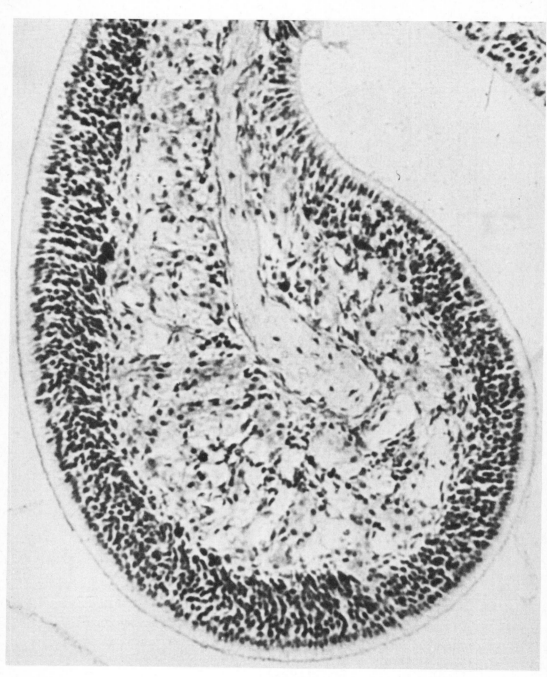

Fig. 8. Migration of dividing cell populations in the olfactory epithelium of mice. The position of labelled cells (darker nuclei) is seen at the various periods after injection of the label (tritiated thymidine).

a) 20 minutes survival period. Labelled cells are concentrated in the basal zone.

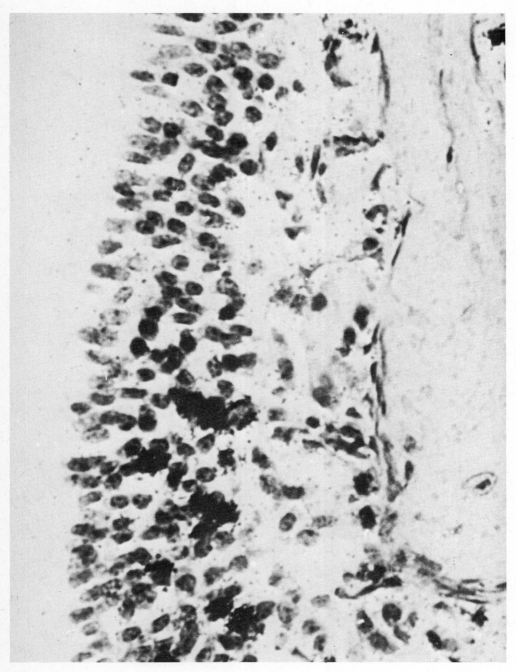

Fig. 8. b) 8 days survival period. A proportion of the labelled cells have migrated peripherally and are in the zone occupied by the nuclei of receptor cells (mid zone).

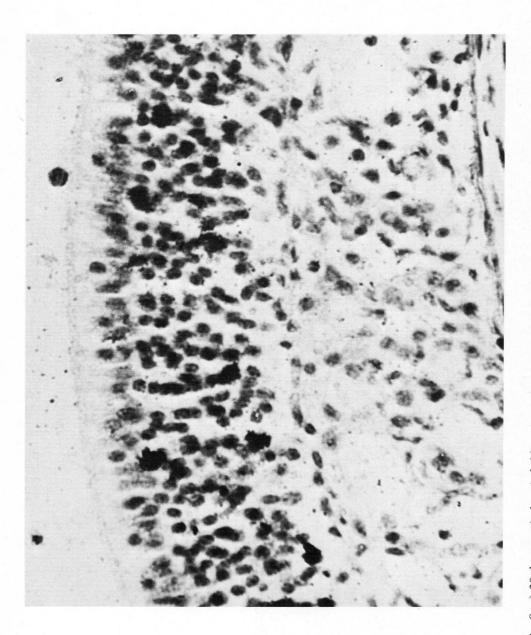

Fig. 8 c) 20 days survival period. Migration is largely completed. The majority of labelled nuclei are now in the mid zone.

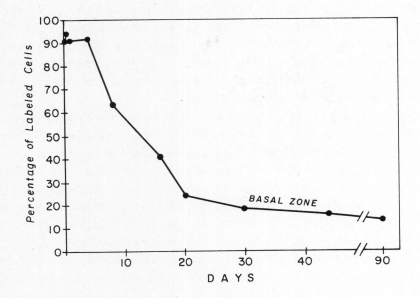

Fig. 9. Migration of labelled nuclei in mouse olfactory epithelium. The
percentage of all labelled cells that are found in the basal zone
are shown for different periods, following the injection of tritiated thymidine. Thus, initially (20 minutes) over 90 % of all labelled nuclei lie in the basal zone. (Data from Moulton et al.,
1970).

(1970), on the basis of histochemical observations, conclude that basal cells do not give rise to
supporting cells. But even the replacement of supporting cells might be expected to disturb the
function of a system in which close structural functional interrelations between cell types are
known to exist (Moulton & Beidler, 1967). It remains to be determined how, in the fact of a
continually shifting cell population, this system maintains constant those features of receptor
response that permit the day-to-day stability of odor recognition.

DIFFERENTIAL SENSITIVITY TO ODORS

In analyzing sensory function a central problem is to determine how receptors differentiate between the particular qualities of a given modality. Perhaps it is simplest to suppose
that for each modality there exists a corresponding receptor type. This was Pfaffmann's (1941)
approach when he first sought the neural basis for salt, sweet, bitter, and sour sensations. When
this is inadequate – as it has been in taste – or impractical – as it is in olfaction, there are several alternatives. One is to assume that there are a limited number of primary qualities, corresponding to which there are a limited number of receptor types. Recognition of quality is then based
on excitation of these receptor types either singly or in various combinations. This idea has worked well for color vision.

If we apply this concept to olfaction we are led to expect that each receptor is especially sensitive to a particular group of odors, and that we should be able to classify odors according to these groupings. But recent single unit recordings from frog, vulture, and tortoise receptors do not seem to bear this out (Gesteland et al., 1965; Lettvin and Gesteland, 1965; Altner and Boeckh, 1967; O'Connell and Mozell, 1969; Shibuya and Tucker, 1967; Mathews and Tucker, 1966). Most units seem to respond to a rather wide range of odors. On the other hand they do not respond to all odors, and different receptors have different spectra of response, so there is a kind of partial specificity. However, it is difficult to see how these responses could be used to classify odors into a relatively small number of types, particularly when the influence of concentration is taken into account.

At the bulbar level, mitral units in a variety of species also respond to many of the odorants tested. Some workers, however, have recorded units that show a higher order of specificity, particularly at lower concentrations, than appears to exist at the receptor level. Adrian (1953), for example, found units that were specifically sensitive to trimethylamine, acetone, ethyl acetate, amyl acetate, pentane, octane, xylol, petrol, clove oil, eucalyptus and other compounds. He considered a unit to show specificity if it was more sensitive to a compound than neighboring units, and reacted at concentrations which had no effect on them. He concluded that there was little chance of finding a relatively small number of primary odors.

In general, most studies of bulbar unit activity have tended to support this conclusion. The few attempts that have been made to derive a classification yield no real agreement about its nature.

At still higher level, in the prepyriform cortex, there is still less evidence of specificity and there is a marked increase in the amount of inhibition observed (Haberly, 1969).

The conclusion which seems to be emerging from this evidence is that odor recognition is the function of a cell ensemble in which individual receptors possess multiple and overlapping sensitivities. Information reaches the brain as a pattern of activity distributed in space in which the ratio of excitation in a number of parallel channels may define the stimulus. To test this possibility it is important to compare activity from a number of sites simultaneously. Successive comparisons are less effective because conditions cannot be duplicated exactly from stimulus-to-stimulus. Some idea of the type of information that this approach can give is shown in Fig. 10. This shows the profiles of excitation derived from simultaneous recording of multiunit activity in the olfactory bulb of the unanesthetized rabbit. These results suggest that different odors can set up different patterns of excitation which may be adequate to characterize a given odor.

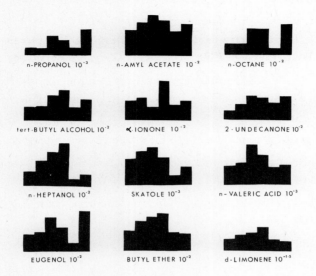

n-PROPANOL 10^{-3} n-AMYL ACETATE 10^{-2} n-OCTANE 10^{-2}

tert-BUTYL ALCOHOL 10^{-2} α-IONONE 10^{-2} 2-UNDECANONE 10^{-2}

n-HEPTANOL 10^{-2} SKATOLE 10^{-3} n-VALERIC ACID 10^{-3}

EUGENOL 10^{-2} BUTYL ETHER 10^{-2} d-LIMONENE 10$^{-1.5}$

Fig. 10. Patterns of odor-induced excitation in the olfactory bulb of the rabbit. Each block of six bars represents the responses of six bulbar sites to a given odor. (The same six sites generated all blocks). Responses of these sites to a given presentation of an odor were recorded simultaneously and the accumulated responses over a series of trials were used to generate the blocks. The response measured was the maximum amplitude of averaged multiunit activity of second order neurones. (From Moulton, D.G. (1970). Differential sensitivity to odors. Cold Spring Harbor Symposia on Quantitative Biology 30, 201-206).

ACKNOWLEDGMENTS

I am greatly indebted to Dr. P.P.C. Graziadei for allowing me to use his photomicrographs and to my students, R.P. Fink and J. Kauer, for their assistance in providing light and electron micrographs respectively.

REFERENCES

Adrian, E.D. 1953. Sensory messages and sensation. The response of the olfactory organ to different smells. *Acta Physiol. Scand. 29*, 4-14.

Adrian, E.D. and Zotterman, Y. 1926. Impulses from a single sensory end organ. *J. Physiol. 61*, viii.

Altner, H. and Boeckh, J. 1967. Ueber das Reaktionsspektrum von Rezeptoren aus der Riechschleimhaut von Wasserfröschen *(Rana esculenta). Z. Vergleich Physiol. 55*, 299-306.

Altner, H. and Müller, W. 1968. Elektrophysiologische und elektronmikroskopische Untersuchungen an der Riechschleimhaut des Jacobsonschen Organs von Eidechsen *(Lacerta). Z. Vergleich Physiol. 60*, 151-155.

Amoore, J.E. 1967. Specific anosmia : A clue to the olfactory code. *Nature 214*, 1095-1098.

Andres, K.H. 1965. Differenzierung und Regeneration von Sinneszellen in der Regio olfactoria. *Naturwissenschaften 17*, 500.

Andres, K.H. 1966. Der Feinbau de Regio olfactoria von Makrosmatikern. *Z. Zellforsch. Mikroskop. Anat. 69*, 140-154.

Andres, K.H. 1968. Neue Befunde zur Feinstruktur des olfactorischen Saumes. *J. Ultrastruct. Res. 25*, 163.

Andres, K.H. 1970. Discussion. In Wolstenholme, G.E.W. and J. Knight (editors). Taste and Smell in Vertebrates. London : J. and A. Churchill, p. 248.

Bang, B.G. and Bang, F.B. 1963. Responses of upper respiratory mucosae to dehydration and infection. *Ann. N.Y. Acad. Sci. 106*, 625-630.

Bannister, L.H. 1965. The fine structure of the olfactory surface of teleostean fishes. *Quart. J. Microscop. Sci. 106*, 333-342.

Bannister, L.H. 1968. Fine structure of the Vomero-nasal organ of the slow worm. *(Anguis fragilis). Nature 217*, 275-276.

Beidler, L.M. and Smallman, R. 1965. Renewal of cells within taste buds. *J. Cell. Biol. 27*, 263-272.

Brightman, M.W. and Reese, T.S. 1969. Junctions between intimately apposed cell membranes in the vertebrate brain. *J. Cell. Biol. 40*, 648-677.

Briggs, M.H. and Duncan, R.B. 1961. Odor receptors. *Nature 191*, 1310-1311.

Brown, H.E. and Beidler, L.M. 1966. The fine structure of the olfactory tissue in the black vulture. *Federation Proc. 25*, 329.

Conger, A.D. and Wells, M.A. 1969. Radiation and aging effect on taste structure and function. *Radiation Res. 37*, 31-49.

Darwin, C. 1875. The variation of plants and animals. New York : Appleton-Century-crofts. 2, 405.

Dieulafé, L. 1906. Morphology and embryology of the nasal fossae of vertebrates. *Ann. Otol. Rhinol. Laryngol. 15*, 267-349.

Garten, S. 1900. Physiologie der marklosen Nerven. G. Fisher, Jena.

Gerebtzoff, M.A. 1953. L'olfaction, structure de l'organe olfactif et mécanisme de l'olfaction. *J. Physiol. (Paris) 45,* 247-283.

Gesteland, R.C., Lettvin, J.Y. and Pitts, W.H. 1965. Chemical transmission in the nose of the frog. *J. Physiol. (Lond.) 181,* 525-559.

Graziadei, P. and Bannister, L.H. 1967. Some observations on the fine structure of the olfactory epithelium in the domestic duck. *Z. Zellforsch. Mikroskop. Anat. 80,* 220-228.

Graziadei, P. and Tucker, D. 1968. Vomero-nasal receptors ultrastructure. *Federation Proc. 27,* 583.

Guillot, M. 1948. Anosmies partielles et odeurs fondamentales. *Compt. Rend. Acad. Sci. 226,* 1307-1309.

Haberly, L.B. 1969. Single unit responses to odor in the prepyriform cortex of the rat. *Brain Research 12,* 48-1484.

Hilding, A. 1932 a. Physiology of drainage of nasal mucus; flow of mucus currents through drainage system of nasal mucosa and its relation to ciliary activity. *Arch. Oto Laryngol. 15,* 92-100.

Hilding, A. 1932 b. Physiology of drainage of nasal mucus III. Experimental work on accessory sinuses. *Am. J. Physiol. 100,* 664-670.

Kauer, J.S. 1969. A study of the structure and function of the vomeronasal (Jacobson's) organ, including investigations into its ultrastructure in mammals. M.A. Thesis. Clark Univ., Worcester, Mass.

Kauer, J.S. and Moulton, D.G. 1970. Ultrastructure of vomeronasal and olfactory neurosensory epithelia in rat and rabbit. *Federation Proc. 29,* 521.

Kendall, D.A. and Neilson, A.J. 1964. Correlation of subjective and objective odor responses. *Ann. N.Y. Acad. Sci. 116,* 567-575.

Kurihara, K. 1967. Isolation of chromoproteins from bovine olfactory tissues. *Biochim. Biophys. Acta 148,* 328-334.

Lettvin, J. and Gesteland, R.C. 1965. Speculations on smell. *Cold Spring Harbor Symposium 30,* 217-225.

Lowenstein, W.R. and Rathkamp, R. 1958. The sites for mechano-electric conversion in a Pacinian corpuscle. *J. Gen. Physiol. 41,* 1245-1265.

Mathews, D.F. and Tucker, D. 1966. Single unit activity in the tortoise olfactory mucosa. *Federation Proc. 25,* 329.

Milas, N.A., Postman, W.M. and Heggie, R. 1939. Evidence for the presence of Vitamin A and carotenoids in the olfactory area of steer. *J. Am. Chem. Soc. 61,* 1929-1930.

Misra, T.N., Rosenberg, B. and Switzer, R. 1968. The effect of adsorption of gases on the semiconductive properties of all-trans-β-carotene. *J. Chem. Physics 48,* 2096-2102.

Moulton, D.G. 1960. Studies in olfactory acuity 5. Comparative olfactory sensitivity of pigmented and albino rats. *Animal Behavior 8,* 129-133.

Moulton, D.G. 1962. Pigment and the olfactory mechanism. *Nature 195,* 1312-1313.

Moulton, D.G. 1965. Differential sensitivity to odors. In Sensory Receptors. *Cold Spring harbor Symposium 30*, 201-206.

Moulton, D.G. 1970. The olfactory pigment. In Beidler, L.M., (editor) Chemoreception. In Handbook of Sensory Physiology. Berlin. Springer-Verlag (in press).

Moulton, D.G. and Beidler, L.M. 1967. Structure and function in the peripheral olfactory system. *Physiol. Revs. 47*, 1-52.

Moulton, D.G., Celebi, G. and Fink, R.F. 1970. Olfaction in mammals – two aspects : proliferation of cells in the olfactory epithelium and sensitivity to odors. In Wolstenholme, G. E.W. and J. Knight (editors). Taste and Smell in Vertebrates. (Ciba Foundation Symposium). London : J. and A. Churchill, 227-246.

Niccolini, P. 1954. Lo stimolo olfattorio et la sua recezione. *Arch. Ital. Sci. Pharmacol. 4*, 109-172.

O'Connell, R.J. and Mozell, M.M. 1969. Quantitative stimulation of frog olfactory receptors. *J. Neurophysiol. 32*, 51-63.

Ogle, W. 1870. Anosmia (or cases illustrating the physiology and pathology of the sense of smell.) *Med. Chir. Trans. 35*, 263-290.

Okano, M., Weber, A.F., Fromms, S.P. 1967. Electron microscopic studies of the distal border of the canine olfactory epithelium. *J. Ultrastructure Res. 17*, 487-502.

Pfaffman, C. 1941. Gustatory afferent impulses. *J. Cellular Comp. Physiol. 17*, 243-258.

Reese, T.S. 1965. Olfactory cilia in the frog. *J. Cell. Biol. 25*, 209-230.

Reese, T.S. and Brightman, M.W. 1970. Olfactory surface and central olfactory connections in some vertebrates. In Knight, J. (editor) Mechanisms of smell and taste in vertebrates. (Ciba Foundation Symposium), Baltimore: Williams and Wilkens (in press).

Rosenberg, B., Misra, T.N. and Switzer, R. 1968. Mechanism of olfactory transduction. *Nature 217*, 423-427.

Shantha, T.R. and Nakajima, Y. 1970. Histological and histochemical studies on the Rhesus monkey *(Macaca Mulatta)* olfactory mucosa. *Z. Zellforsch. Mikroscop. Anat. 103*, 291-319.

Shibuya, T.S. and Tucker, D. 1967. Single unit responses of olfactory receptors in vultures. In Hayashi, T. (editor) Olfaction and Taste 2, Oxford : Pergamon Press 219-233.

Steinbrecht, R.A. 1969. Comparative morphology of olfactory receptors. In Pfaffmann, C. (editor). Olfaction and Taste III 1-21. New York : Rockefeller University Press.

Stoll, M. 1965. De l'effet important de différences chimiques minimes sur la perception de l'odeur. *France et ses Parfums 8*, 227-232.

Stuiver, M. 1958. Biophysics of the sense of smell. Doctoral Thesis. Rijks Univ. Groningen, The Netherlands.

Thornhill, R.A. 1967. The ultrastructure of the olfactory epithelium in the lamprey. *(Lampetra fluviatilis). J. Cell. Sci. 2*, 59-1602.

Tucker, D. 1963. Olfactory, vomero-nasal, and trigeminal receptor responses to odorants. In Olfaction and Taste. Y. Zotterman, Ed., 45-69. Oxford : Pergamon Press.

Tucker, D. 1967. Olfactory cilia are not required for receptor function. *Federation Proc. 26*, 544.

Wilson, J.A.F. and Westerman, R.A. 1967. The fine structure of the olfactory mucosa and nerve in the teleost *Carassius carassius* L. *Z. Zellforsch. Mikroscop. Anat. 83*, 196-206.

Wright, R.H., Reid, C. and Evans, H.G.V. 1956. Odor and molecular vibration (3). A new theory ol olfactory stimulation. *Chem. and Industry 37*, 973-977.

DISCUSSION

Dravnieks : Although a tendency exists to relate odorant recognition to interaction with some as yet unknown receptor substance, evidence is accumulating that common physicochemical properties of odorants may account for much of the odorant discrimination. Laffort (Paris) has proposed expressing these properties through a set of three numerical parameters : a (related to molvolume), π (related to hydrogen bonding ability), and ϵ (related to the electronic polarizability of a part of the odorant molecule). These parameters can be considered as three coordinates. Recently, Laffort has shown that physicochemical similarity between odorants, if expressed in terms of proximity between the two points representing respective odorants in the a, π, ϵ space, correlates significantly ($p < 0.001$) with (1) psychophysical similarity in humans as expressed by Woskow's (1968) data, and (2) electro-physiological similarity at mitral cell level in frogs, as expressed by Døving's (1966) data. This tends to support the significance of common properties of odorants in recognition of similarity which is the corner stone of odor discrimination. It is not expected, of course, that all discrimination is based on these common factors only. Special parameters, perhaps steric topology of the odorant molecules in the sense of Beets profile functional group theory, probably enter as significant factors, especially in pheromone olfaction.

Zottermann : What does "renewal of olfactory sense" imply? This appears to be a new finding.

Kalmus : We have found that a subject who lost his sense of taste after accidental irradiation recovered apparently normally after a few weeks. The evidence was that there was a loss of tissues, but the cells were then regenerated.

Zottermann : But the problem here is the olfactory system.

Kalmus : Has the olfactory epithelium been histologically investigated after oblation?

Moulton : (in reply to Kalmus). Schultz (1960) reports regeneration in the olfactory epithelium of the Rhesus monkey following zinc sulphate necrosis. However, it required six months and no tests for functional normalcy were applied. Other instances have been reported, chiefly in non-mammalian forms (see Moulton and Beidler, 1967; Moulton et al., 1970).

Schneider : There is plenty of evidence for regeneration of nerve cells and fibers in the olfactory system, as Andres has shown.

Steinbrecht : Andres has suggested that not only is the whole olfactory receptor cell regenerated, but that the outer peripheral parts of the receptor cells, which are much more prone to damage, are replaced more rapidly.

Moulton :   I did not say that regeneration of olfactory receptors does not occur and I am fully aware of precedents for regeneration following experimental interruption of neurones in sympathetic, visual and other systems. What I was concerned with was not regeneration, or continuous growth, but continuous cell renewal (in the adult) not experimentally or pathologically induced. Although related, these are separate and clearly defined phenomena with different functional implications (see, for example, Leblond and Walker, 1956). A simple demonstration that mitosis occurs is not necessarily a sufficient basis for identifying a renewing population. What is needed is quantitative evidence concerning the relative proportions of dividing and non-dividing cells and their movements over a significant period of time. In this context the idea that olfactory receptors undergo continuous renewal is not well established. Thus Andres (1970) appears to have withdrawn his earlier suggestion that olfactory receptors undergo continuous renewal in so far as he finds mitosis to be rare in adults. He also reports that the "moulting" and replacement of receptor dendrites does not occur in the rat and he has suggested that where it does occur it may be induced by alterations resulting from a virus disease.

Wright :   The olfactory cilia are one micron diameter and one to three hundred long. It is difficult to imagine any chemical transformation going on in a pipe of those dimensions, so the primary process of stimulation must be a physical one not involving making or breaking of chemical bonds.

Kaissling :   We know from insect sensory cells that they react to single molecule impacts, possibly by making a hole that allows ions to pass.

Schneider :   We do not have evidence about this. Perhaps the cilia are molecule-catching devices channelling the molecules to an area where there is a true membrane which then can carry out the transduction.

REFERENCES TO DISCUSSION

Andres, K.H. 1965. Differenzierung und Regeneration von Sinneszellen in der Regio olfactoria. *Naturwissenschaften, 17,* 500.

Andres, K.H. 1970. In Discussion following paper by Moulton, et al. In Wolstenholme G.E.W., and Knight, J. (editors) Taste and Smell in Vertebrates. Churchill : London. 247-248.

Døving, K.B. 1966. An electrophysiological study of odor similarities of homologous substances. *J. Physiol. 186,* 97-109.

Moulton, D.G. and Beidler, L.M. 1967. Structure and Function in the Peripheral Olfactory system. *Physiol. Revs. 47,* 1-52.

Moulton, D.G., Celebi, G. and Fink, R.P. 1970. Olfaction in mammals – two aspects : proliferation of cells in the olfactory epithelium and sensitivity to odors. In Wolstenholme, G.E.W., and Knight, J. (editors). Taste and smell in vertebrates. 227-245, Churchill : London.

Leblond, C.P. and Walker, B.E. 1956. Renewal of cell population. *Physiol. Rev. 36,* 255-275.

Schultz, E.W. 1960. Repair of the olfactory mucosa with special reference to regeneration of olfactory cells (sensory neurones). *Am. J. Pathol. 37,* 1-19.

Woskow, M.H. 1968. In Theories of Odor and Odor Measurement, Tanyolac ed., Robert College Istanbul, p. 147.

AN EXPERIMENTAL APPROACH TO THE PERIPHERAL MECHANISMS

OF OLFACTORY DISCRIMINATION

P. Mac Leod

**Laboratoire de Neurophysiologie du Collège de France,
11 Place Marcelin Berthelot, Paris 5°.**

INTRODUCTION

The olfactory receptor is a transducing organ that provides the link between the external environment where the odor molecules are present and the brain which recognizes their presence.

The whole complex of olfactory receptors forms a coding system that generates and transmits information relating to a physico-chemical phenomenon, the molecule/receptor interaction, in the form of electrical impulses.

Thanks to the large body of experimental data that has recently been built up, we have acquired a very good knowledge of this system; chemistry tells us a great deal about the structure and properties of the stimulating molecules; the receptors have gradually revealed to us the minutest details of their anatomy; electrophysiology and psychophysiology enable us to detect their response with increasingly greater accuracy. But one essential item of information is lacking : we do not yet know how to decipher their message whose code we have still not cracked.

Between the molecules – about which we know practically everything, and the receptors– about which there is not much left to discover, one essential link is missing. We just cannot "see" exactly what happens between the moment the molecules are adsorbed on the membrane of the olfactory cilia and the beginning of the local electrical response. If we could, we should have no trouble decoding the olfactory message and our investigation would be completed.

All that we can do to get round this last difficulty is to try and guess the course and dimensions of this invisible interaction by interpolating what we know of both olfactory information and odorous molecules, adopting a procedure analogous to that of the cryptologist deciphering a bilingual inscription.

Although simple in conception, in practice such a procedure comes up against the problem of the enormous number of points of intersection to be considered; several tens of millions of independent olfactory receptors work together to identify several tens of thousands of different odorous stimuli.

The means of data gathering and processing we have at present at our disposal rule out the radical solution which would consist in examining all possible cases, i.e. describing and indexing the nerve messages corresponding to all odors. We are thus obliged to adopt an intermediate solution which consists in working on the basis of limited samples. This procedure has the advantage of being a perfectly objective one and avoids to the greatest possible extent any interference between the experimental results and the experimenter's own prejudices. It also has the obvious disadvantage that it is tedious and requires the application of techniques taken to the uttermost limits of their possibilities. It seems at present to be the most likely to enable us to pass beyond the first stage at the end of which the most fruitful lines of research will be defined, these lines then being framed in the form of working hypotheses based solidly on facts and not on pure speculations. The procedure is now in process of giving us a definitive answer to an essential preliminary question : how many dimensions define the olfactory phenomenon? It will probably enable us subsequently to find out the exact nature of these dimensions.

In order to take stock of the present situation, we must now review the experimental data we can henceforth rely upon. We can classify them under two heads according to the techniques involved in their acquisition : morphology and electrophysiology.

MORPHOLOGICAL DATA

The olfactory chemoreceptor, a bipolar neurone differentiated into a sense receptor, is of particular interest to us in respect of its dendritic extremity, the only part which comes into contact with the olfactory stimuli. The olfactory cilia emerging from the olfactory vesicle form a very dense mat inside the layer of mucus covering the neuroepithelium, and the surface factor thus obtained may be estimated to be 100, i.e. 1 cm^2 of neuroepithelium corresponds to 100 cm^2 of effective contact surface between the mucus and the chemoreceptive membranes (Moulton and Beidler, 1967). In view of the dimensions of its meshes and the mean amplitude of brownian movement in the mucus at normal temperature, this device constitutes an excellent molecule trap.

The olfactory cilia are fragile and perpetually being renewed; basal corpuscles are constantly being synthesized in the perinuclear zone, they migrate towards the periphery and, when they arrive in olfactory vesicle, they bring about the formation of nine pairs of filaments which, extended, evaginate the membrane to form a new cilium (Helst and Mulvaney, 1968). In addition to these processes of cilia formation, many degenerative aspects are observed – localized distensions, thinning of the ends, some broken off (Reese, 1965).

The cilia are mobile, but their movements, in contrast to those of the respiratory cilia, are not synchronized (Tucker 1967; Graziadai, 1969; Unpublished results).

The olfactory receptors occasionally exhibit localized coupling with thinning of the intervening membrane which loses its double membrane character. Up to now, no true cytoplasmic bridges have been observed, and the functional significance of these couplings is still unknown. (Frisch, 1967; Graziadei, 1966).

The axonic extremity of the olfactory receptor, which has been studied very thoroughly by Casser (1956) and Lorenzo (1957) does not exhibit any peculiarity as compared with the normal arrangement amyelinic nerve fibres with schwannian collective sheath. No experimental findings have substantiated the idea that the close intermembrane spacing might favour electrical coupling between adjacent fibres. Calculations show, on the contrary, that 200 Å of extracellular conducting space is more than enough to ensure the independence of nerve messages (Ottoson, 1967).

The termination of the olfactory axones in the olfactory glomerules is characterized by an extraordinary profusion of ramifications and synaptic junctions (de Lorenzo, 1963; Reese, 1969) to the extent that the sensory information conveyed by each afferent fibre is distributed to all the efferent fibres of a glomerule. Centrifugal synaptic junctions have been described in connection with the dendrites of the mitral cells, but the latter do not concern the olfactory fibres; they are dendro-dendritic, probably with the dendrites of the interglomerular interneurones (Reese, 1969; Altner, 1969).

Consequently, there is nothing to invalidate the concept of the mutual independence of the olfactory fibres and of the information they convey. The complex of feedback circuits whose various loops help to define precisely and to stabilize the olfactory message is closed inside the olfactory bulb and no centrifugal connection liable to modify the sensitivity of the receptors has yet been disclosed.

ELECTROPHYSIOLOGICAL DATA

The electrical response of the olfactory receptors was first recorded indirectly in the form of a slow, predominantly negative variation in potential at the surface of the neuroepithelium with an amplitude of several millivolts and a duration of several seconds (Hosoya and Yoshida, 1937; Ottoson, 1956; Mac Leod, 1959). It was given the name of electroolfactogram (E.O.G.).

The E.O.G. is easily recorded without disturbing the functioning of the olfactory receptors. It is thus especially convenient when dealing with problems of olfactory discrimination.

After a considerable amount of controversy, the recent work of Gesteland (1963) and Getchell (1969) has clearly shown the extent to which electrochemical artefacts can be added to the generator potential and how to avoid them.

As long as certain precautions are taken, therefore, the E.O.G. can be used to give an overall picture of the state of activation of a huge population of olfactory receptors. It was by using the E.O.G. that it was possible to see for the first time that the electrical response of the olfactory receptors is not solely and invariably negative and that some receptors respond to certain stimuli by becoming hyperpolarized. It was also the E.O.G. that enabled Getchell to demonstrate that certain odorless substances such as urea, glucose or sucrose are just as good olfactory stimuli as ordinary odorous substances provided they are brought into contact with the mucosa in aqueous solution : their lack of odor in ordinary conditions is due merely to their low volatility. We thus are brought round to the concept that the olfactory chemoreceptors are truly universal detectors capable of responding to all molecules that can be adsorbed in sufficient numbers on the surface of their cilia.

With reference to the problems of discrimination, the possibilities of E.O.G. analysis are restricted to the study of its decay with time and its variations as a function of the nature of the stimulus or the interactions between different stimuli; it is hardly possible to derive from it any information about the topographical distribution of the receptor activity. Now, it seems that for olfaction, as for other sensory processes, qualitative discrimination depends chiefly on the analysis of this distribution by the nerve centres. As any item of sensory information is ultimately reflected in a space/time distribution of electrical activity in the afferent nerve fibres, and as the number of fibres is infinitely greater than the number of available "instants" on which to base a qualitative judgment, it is quite obvious that, for the experimenter, the space dimension is the essential thing he has to find out.

This is what justifies the very many attempts made to record the unit response of the olfactory receptors, ranging from the activity of the olfactory fibres recorded by juxtacellular metal micro-electrodes (Gesteland, 1963; Shibuya, 1963; O'Connell and Mozell, 1969), to the unit receptor potential recorded by an ultrafine intracellular micro-electrode (Aoki, 1968) – a feat which still seems to be a long way from becoming routine practice.

Unfortunately, although their strictness is in itself satisfying, these unit recordings have proved disappointing from the point of view which interests us, as it is impossible to make enough of them simultaneously on the same preparation to give even a rough outline of the topography of the activation. Nonetheless, on the credit side, they have given us an accurate idea of the selectivity of the individual receptors, which appears to be very low and at present impossible to systematize, and they have confirmed the existence of two types of response, activation and inhibition, each receptor being capable of either – or none at all – depending on the nature of the stimulus. O'Connell's and Mozell's quantitative measurements have, moreover, showed that, on the quantitative level, Stevens's (1957) power law applies to olfactory receptors considered individually and in the case of concentrations lower than those producing the maximum response.

If we are to move a step nearer to understanding the olfactory transduction code, it seems that we are obliged to take the greatest possible sample of elementary responses produced by the greatest possible number of different stimuli and to look, in the matrix thus obtained, for the existence of correlations between odors or between receptors; it might then be possible, by examining these correlations, to demonstrate the existence of both receptor processes common to several receptors and stimulating processes common to several molecules.

If such a procedure is to be employed, each unit studied must remain fairly constant over several hours, for olfactory stimulation cannot be repeated at intervals of less than two or three minutes without risk of the responses being distorted by adaptation phenomena. Recording the unit responses of olfactory fibres has not proved very practical up to now. We have therefore attempted to exploit two experimental artifices : recording firstly very highly localized E.O.G.s and, secondly, unit responses of olfactory glomerules.

Thanks to the assistance of J. Leveteau, we were able to investigate the selectivity of almost 200 olfactory glomerules of the rabbit towards 12 odorous substances that were as different from each other as possible on both the chemical and human sensory levels (Leveteau and Mac Leod, 1966, 1969). The results we obtained were both well defined and inconclusive. They were well defined in the sense that each glomerule responds distinctly to some of the odors presented to it and not at all to the rest; this clear-cut distinction made it possible to represent their discrimination by a binary matrix in which three spaces out of five are filled but in which each of the 12 lines corresponding to the various stimuli represents a quite distinct sequence (Fig. 1). Similarly, the columns corresponding to the response profiles of the individual glomerules are almost all different and the few identical pairs could doubtless have been differentiated by using one or two additional stimuli. Although visual inspection reveals no systematization in these data, calculating the correlation coefficient χ^2 by pairs between the 12 stimuli shows that some of the 12 substances selected unquestionably exhibit common stimulatory actions (Fig. 2).

These similarities have been presented in graphical form by Døving (1970) on the basis of a non-parametric analysis of the whole set of χ^2 coefficients ranked in descending order, in two forms : firstly a tree diagram (hierarchy clustering scaling), and secondly a multidimensional presentation (multidimensional scaling). In both cases these are proximity diagrams in which the distance separating two points representing two odors is all the smaller the greater the degree of similarity between these two odors. The second form of presentation does, however, supply an additional item of information : the number of dimensions required to obtain a coherent diagram. In the present case, Døving showed that at least four were needed. This means, all in all, that the qualitative olfactory message is defined by at least four independent parameters, but, of course, it gives us no information at all on the nature of these parameters.

Fig. 1. Table of unit glomerule responses.
        96 glomerules were tested, with 12 different stimuli; a black square
        corresponds to a response, a white square to the absence of a response.

A : Methanol	B : Propanol	C : Naphthalene
D : Coumarin	E : Pyridine	F : Nitrobenzene
G : Camphor	H : Benzene	I : Heptanol
J : Isoamyl acetate	K : Decanol	L : β-Ionone

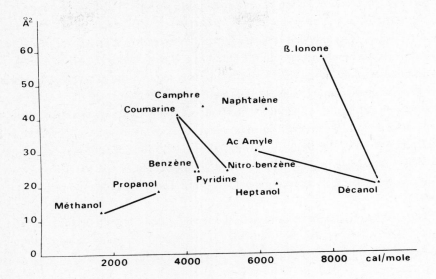

Fig. 2. Correlations by pairs according to the calculation of χ^2. The stimuli in Fig.1 are presented on a two-dimensional diagram. x-Axis : the heat of adsorption; y-axis : cross-section of the molecule (after Davies). The solid lines show the most significantly correlated pairs (p $\leqslant$ 0.01).

Continuing on the same experimental lines, with the collaboration of J. Leveteau and G. Daval, we endeavoured to make use of the drop in the electrical conductivity of the olfactory mucus to obtain localized E.O.G.s in frogs. We were encouraged to do this by the work of Tucker and Shibuya (1965) who demonstrated the possibility of modifying the ion environment of the olfactory receptors without interfering with their functioning, and by those of Døving and Mustaparta (1969) who showed that different points of the olfactory epithelium of the frog rank the amplitudes of the E.O.G.s produced by the same series of stimuli in different orders. Our technique consists in irrigating the nasal cavity with an isotonic sucrose solution and in presenting the stimuli directly in an aqueous phase. In these conditions, the E.O.G. obtained by a micropipette one to two microns in diameter placed in contact with the neuroepithelium corresponds to the activity of a zone of mucosa limited to a radius of 100 to 150 microns around the point of contact of the pickup electrode. The results obtained show that the relative amplitudes of the different E.O.G.s produced by the 12 substances already employed in studying the glomerules of rabbit vary considerably from one point to another. Just as in the case of the glomerules it was possible to calculate the χ^2 correlation coefficients between pairs of stimuli, here the calculation of Pearson's correlation coefficients showed that certain substances are more closely linked, which suggests the existence of common stimulatory processes.

A method of statistical analysis recently developed by Benzecri (1969), factor analysis of correspondences (analyse factorielle des correspondances) was applied to our whole set of results. This method belongs to the category of proximity analyses. Like all factor analyses, it starts out from a table of numerical data built up from juxtaposed equal columns or rows. In the case under consideration, each row corresponds to a given receptive structure and each column to a given odor stimulus. At the intersection of each column and row we find the measurement of the response of the receptive structure represented by this row to the odor stimulus represented by this column. Its original feature is that it makes it possible to construct, in a space of several dimensions, a set of points representing both the columns and the rows of the original table.

In these conditions, proximity means similarity, so that the existence of correspondences between certain elements in this set gives rise to identifiable sub-sets. These regroupings are particularly significant, for they simultaneously reflect similarities between elements in the same initial sub-set (columns **or** rows) and correspondences between elements belonging to both initial sub-sets (columns **and** rows). Reducing the initial space to a sub-space of minimum dimensions additionally makes it possible to determine how many independent factors are necessary and sufficient to describe correctly the observed phenomena : just as many factors as dimensions in the reduced space.

Three series of tests carried out with the same set of 12 stimuli on three different preparations – the olfactory glomerules of rabbit (Fig. 3), the olfactory epithelium of rabbit (Fig. 4), and the olfactory epithelium of frog (Fig. 5) – were subjected to factor analysis of correspondences.

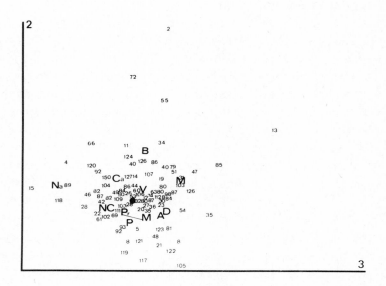

Fig. 3. Factor analysis of correspondences : unit glomerulus responses. Analysis of the data in Fig.1 supplemented by an additional series of tests. The points have been shown projected onto the plane formed by the second and third axes (y and x-axis respectively).

The figures represent the glomerules, the letters refer to the stimuli : A : Isoamyl acetate; B : Benzene; C : Coumarin; Ca : Camphor; D : Decanol; H : Heptanol; M : Methanol; N : Nitrobenzene; Na : Naphthalene; P : Pyridine; Pr : Propanol; V : β-Ionone.

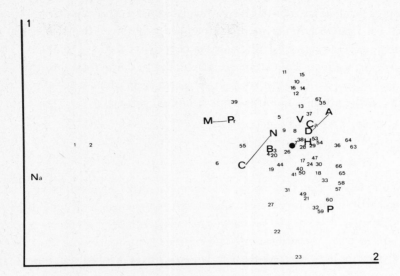

Fig. 4. Factor analysis of correspondences : E.O.G. of rabbit.

Same conventions as for Fig. 3 : the figures represent different points at which the E.O.G. was recorded with microelectrodes. The points have been shown projected onto the plane formed by the first and second axes (y and x-axis respectively).

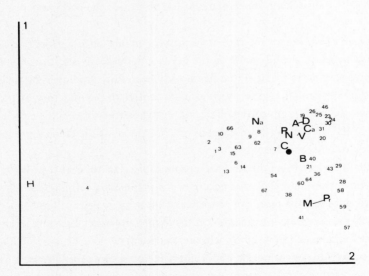

Fig. 5. Factor analysis of correspondences : local E.O.G. of frog. Same conventions as for Figs. 3 and 4.

The dense groups of points obtained are still correctly represented in a four-dimensional space. Some relationships are constantly revealed between the same stimuli in the different preparations : methanol and propanaol; decanol and isoamyl acetate; coumarin, nitrobenzene and pyridine. Moreover, the points corresponding to the receptive structures belonging to the same animal tend to be grouped, which reflects the individual differences in receptivity.

However, the most definite conclusion that can be drawn from this analysis is that it will be necessary to carry out the same type of experiment again with a much greater number of odor stimuli in order to obtain sufficiently dense significant groupings.

DISCUSSION

It is to be regretted that the various research projects at present being conducted on the various aspects of the olfactory system concern different animal species and, in particular, different series of stimuli. It seems that several research workers have now become aware of this and that we shall witness a better coordination of their efforts in the near future. In the present situation, it is impossible for us to compare the correlations we have revealed with those found on the basis of unit responses of neurons in the olfactory bulb or, again, of psychophysical investigations (Døving, 1966, Higashino et al. 1969, Woskow, 1968), whereas it has been possible to establish a number of common similarities between these last two categories of data. At the present time we possess only one indirect element of comparison which is provided by the work of Laffort (1968, 1969). This author has shown that it is possible to calculate the psychophysical parameters of threshold and olfactory intensity for about 50 substances on the basis of measurements of a few physico-chemical quantities involved in the adsorption phenomena. Comparison of the degrees of similarity observed by various workers with those that can be deduced from these physico-chemical measurements shows that the latter give satisfactorily close results as far as the bulb units and the psychophysical phenomena are concerned, but not as regards the peripheral responses (Laffort, 1970).

This lack of agreement should not surprise us unduly since Laffort's calculations are primarily intended to set up a valid model on the psychophysical level and the responses of the bulb neurons represent a nerve message that has already been fully developed by the interplay of all the centrifugal control circuits linking various brain centres to the olfactory bulb. The information collected at the mitral cells of the olfactory bulb therefore reflects the activity of the rest of the brain, i.e. of the subject as a whole, just as much, at least, as the crude sense information coming from the peripheral receptors.

Finally, the only thing that has been proved for certain is that the olfactory process has at least four dimensions, but the exact nature of the parameters defining both the odor properties of the molecules and the receptive processes of the sensory cells will remain open to discussion until we possess accurate information on the nature and exact structure of the proteins with which the membranes of the receptor cells are equipped.

The time factor might not be important if the odor of a substance were perceived and identified from the beginning of the peripheral response; however, this concept is not based on precise experimental data except as regards insects (Oethier, 1968) and it seems that it will be seriously called into question by the present development of Moulton's and Gesteland's researches into the rôle of the time pattern of bulb and peripheral excitation in olfactory perception.

REFERENCES

Andres, K.M. 1970. Anatomy and ultrastructure of the olfactory bulb in fish, amphibians, reptiles, birds and mammals. In : Mechanisms of taste and smell in vertebrates, Ciba, O.E. Lowenstein ed., London, Churchill.

Aoki, K. 1968. Intracellular recording of the olfactory cell activity. *Proc. Jap. Acad. 44,* 856-857.

Benzecri, J.P. 1969. L'analyse factorielle des correspondances. Laboratoire de Statistiques de la Faculté des Sciences de Paris (publication multigraphiée).

Dethier, V.G. 1968. Chemosensory input and taste discrimination on the blowfly. *Science 161,* 389-391.

Døving, K.B. 1966. An electrophysiological study of odor similarities of homologous substances. *J. Physiol. London, 186,* 97-109.

Døving, K.B. 1970. Experiments in olfaction. In :Mechanisms of taste and smell in vertebrates, Ciba, O.E. Lowenstein ed., London, Churchill.

Frisch, D. 1967. Ultrastructure of mouse olfactory mucosa. *Am. J. Anat. 121,* 87-119.

Gasser, H.S. 1956. Olfactory nerve fibers. *J. gen. Physiol. 39,* 473-496.

Gesteland, R.C., Lettvin, J.Y., Pitts, W.H. and Rojas, A. 1963. Odor specificities of the frog's olfactory receptors. In : Olfaction and Taste I, Y. Zotterman ed., Oxford, Pergamon Press, 19-34.

Getchell, T.V. 1969. The interaction of the peripheral olfactory system with non odorous stimuli. In : Olfaction and taste III, C. Pfaffmann ed., New York, Rockefeller University Press, 117-124.

Graziadei, P. 1966. Electron microscopic observation of the olfactory mucosa of the mole. *Proc. Zool. Soc. London, 149,* 89-94.

Heist, H.E. and Mulvaney, B.D. 1968. Centriole migration. *J. Ultrastructure Res. 24,* 86-101.

Higashino S., Takeuchi, H. and Amoore, J.E. 1969. Mechanism of olfactory discrimination in the olfactory bulb of the bullfrog. In : Olfaction and Taste III, C. Pfaffmann ed., New York, Rockefeller University Press, 192-211.

Hosoya, Y. and Yoshida, H. 1937. Ueber die bioelektrische Erscheinungen an der Riechschleimhaut. *Japan. J. Med. Sci. III Biophys. 5,* 22-23.

Laffort, P. 1968. Some new data on the physico-chemical determinants of the relative effectiveness of odorants. In : Theories of odor and odor measurement, N. Tanyolac ed., Maidenhead, Technivision, 247-269.

Laffort, P. 1969. A linear relationship between olfactory effectiveness and identified molecular characteristics. In : Olfaction and Taste III, C. Pfaffmann ed., New York, Rockefeller University Press, 150-157.

Leveteau, J. and Mac Leod, P. 1966. La discrimination des odeurs par les glomérules olfactifs du lapin (étude électrophysiologique). *J. Physiol. Paris, 58,* 717-729.

Leveteau, J. and Mac Leod, P. 1969. La discrimination des odeurs par les glomérules olfactifs du lapin (influence de la concentration du stimulus). *J. Physiol. Paris, 61,* 5-16.

Lorenzo, A.J. de, 1957. Electron microscopic observations of the olfactory mucosa and olfactory nerve. *J. Biophys. Biochem. Cytol. 3,* 839-850.

Lorenzo, A.J. de, 1963. Studies on the ultrastructure and histophysiology of cell membranes, nerve fibers and synaptic junctions in chemoreceptors. In : Olfaction and Taste I, Y. Zotterman ed., Oxford, Pergamon Press, 5-17.

Mac Leod, P. 1959. Premières données sur l'électroolfactogramme du lapin. *J. Physiol. Paris, 51,* 86-92.

Moulton, D.G. and Beidler, L.M. 1967. Structure and function in the peripheral olfactory system. *Physiol. Rev. 47,* 1-52.

Mustaparta, H. 1969. Simultaneous recordings of receptor potentials, electro-olfactograms, from two different sites in the olfactory epithelium in frog. 6e Symposium Méditerranéen sur l'Odorat, Grasse, Monte-Carlo (données non publiées).

O'Connell, R.J. and Mozell, M.M. 1969. Quantitative stimulation of frog olfactory receptors. *J. Neurophysiol. 32,* 51-63.

Ottoson, D. 1956. Analysis of the electrical activity of the olfactory epithelium. *Acta Physiol. Scand. 35,* suppl. 122, 1-83.

Ottoson, D. 1967. Experiments and concepts in olfactory physiology. In : Progress in Brain Research, vol. 23, Sensory mechanisms, Y. Zotterman ed., Amsterdam, Elsevier, 83-138.

Reese, T.S. 1965. Olfactory cilia in the frog. *J. Cell. Biol. 25,* 209-230.

Reese, T.S. 1969. Types of olfactory receptors in vertebrates. In : Mechanisms of taste and smell in vertebrates, Ciba, O.E. Lowenstein ed., London, Churchill, (in press).

Shibuya, T. and Shibuya, S. 1963. Olfactory epithelium : unitary responses in the tortoise. *Science 140,* 495-496.

Stevens, S.S. 1957. On the psychophysical law. *Psychol. rev. 64,* 153-181.

Tucker, S. and Shibuya, T. 1965. A physiological and pharmacological study of olfactory receptors. Cold Spring Harbor Symposium on Quantitative Biology, 30, 207-215.

Woskow, M.H. 1968. Multidimensional scaling of odors. In : Theories of odor and odor measurements, N. Tanyolac ed., Maidenhead, Technivision, 147-188.

DISCUSSION

Hughes :   During hypophysectomy operations on man, we have been able to implant electrodes in the olfactory bulb and record the electrophysiological activity produced when the patient inspired a certain odorant. This type of activity possesses a complex wave form of a sinusoidal type, but the frequency components must be analyzed in a special detailed way. We have tape-recorded the responses, and, after analyzing the various frequency components we have found that odors that are judged by the patient to be in a given odor category, e.g. floral, tend to have similar frequency components.

Recently we have extended our studies to the human amygdala with Dr. Orlando Andy of the University of Mississippi; this structure is situated deep in the temporal lobe, where the final analysis occurs. Electrodes were placed in the amygdala of patients undergoing therapeutic operations for intractable seizures (amygdalotomy), and we recorded the electrophysiological activity from these electrodes while the patient, who is conscious throughout, inspires a given odorant.

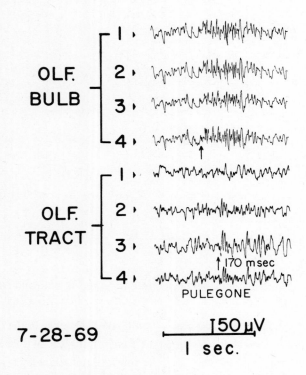

Fig. 1. Electrophysiological activity of olfactory bulb and
olfactory tract.

This figure shows on the top 4 channels the electrophysiological activity from the olfactory bulb, and on the bottom 4 channels from the cut contra-lateral olfactory tract, indicating the activity that is fed back to the opposite bulb. Note the arrow under channel 4, indicating the onset of rhythmical activity in the bulb and then the latency of 170 milliseconds before the onset of activity in the tract.

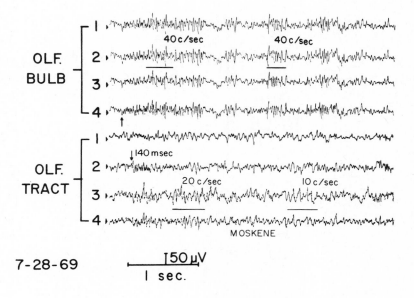

Fig. 2. Latency between onset of rythms in bulb and tract.

There is a latency difference (140 milliseconds) between the onset of the rhythms in the bulb, compared to the tract suggesting that there is quite a long time required for the processing of the information to be fed back into the contralateral olfactory tract. One sees in the bulb frequencies of about 40 per second, and in the opposite olfactory tract frequencies that appear to be subharmonics of the bulb frequencies, at about 10 to 20 per second.

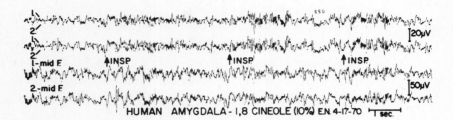

Fig. 3. Rythmical activity for human amygdala in response to 1,8-cineol.

This figure shows rhythmical activity from the human amygdala in response to 1,8 cineol (10%). Note the sinusoidal activity after each inspiration, indicated by the arrows. The first 2 channels record bipolarly between 2 different electrodes in the amygdala and the last 2 channels are referential recordings to the mid frontal area.

Thus, we have shown that the major electrophysiological response to odorants in the olfactory bulb, olfactory tract and amygdala in man is rhythmical, sinusoidal activity. The analysis of these rhythms reveal distinctive frequency components according to the stimulus and, therefore, we believe that these components are directly involved in the neurophysiological coding of information in the olfactory system.

Mac Leod : These frequencies give an image of the fiber processes that is too indirect. The fibers are topographically perfectly defined, but transmit individual frequencies that are badly defined and even random. Sinusoidal rhythms happen as a result of the simultaneous activity of a huge number of specifically excited fibers and can give no information about their topography.

Zottermann : What happens if you raise the intensity of the stimulus, Dr. Hughes?

Hughes : The frequency of the recorded response does not change, but its amplitude decreases.

Harper : How many independent stimuli must be considered in your method, Prof. MacLeod?

Mac Leod : I estimate it at several hundreds. If the different possible groupings are more than about 10, it will need more than 10 constituents to identify the smallest group. Using multi-dimensional methods, the number of independent parameters allowing the interpretation of a probability diagram does not seem to increase when the number of stimuli is increased. The same number of dimensions are required for six stimuli as for twelve, and, I hope, for several dozen.

Tucker :   Did you have calcium in the solutions of the odorous products used by perfusion?

Mac Leod :   Yes.

Wright :   We analyzed the frequencies from a tape-recording of the olfactory brain of the desert locust, which reacts to the odor of freshly cut grass. We first recorded the pattern of response when the antenna were exposed to the odor of freshly cut grass. We then exposed the antenna to a number of odoriferous chemicals and obtained a different pattern of response. It was possible to select three of these substances that, when applied in combination to the insect resulted in behavior that was qualitatively the same as upon exposure to grass, but when applied individually were without action. However, the locust will behave in this way to almost any odor of vegetation, so this result may be fortuitous and for various reasons the work was not continued any further.

Dravnieks :   In a discussion of factors significant in odorant discrimination, the question frequently arises as to how many physicochemical factors manifest themselves in interactions of odorant molecules with condensed phases. Recently, we had an opportunity to statistically factor-analyze the behavior of over 60 compounds in interaction with many condensed phases exemplified by gas-chromatographic retention characteristics. Tentatively, the conclusion is that not more than 8 factors emerge. Two or three of these were rather general, while the rest were special, needed to bring some specific compound class into accord with the general factor matrix. Thus, it appears that the language of intermolecular interactions, by which the odorant communicates with the receptor, operates with not much more than eight factors.

Mac Leod :   Are the eight factors found by Prof. Dravnieks properties of the stationary phases of the columns, or of the 60 chemical species tested?

Dravnieks :   As I just mentioned, these are the factors involved in the **interactions** occurring between chemicals **and** columns. More generally speaking, I mean every intermolecular interactions.

Schneider :   Concerning Dr. Wright's observation, I believe that he is high on the neuron level and in the brain, where motor patterns are being prepared. The difficulty consists in recording impulses of individual fibers of the neuro-pile, which are hardly more than one or two tenths of a micron in diameter.

Kovats :   Are the curves produced by the same stimulus reproducible with the same preparation, and are they also with different preparations?

Mac Leod :   They are completely reproducible on the same preparations and quite well reproducible from one preparation to another, but only on condition that the fiber in question shows a response to the given stimulus.

Kalmus :   I doubt whether these olfactory analyses will lead anywhere. Everything is much more complex than cryptography, mentioned by somebody in this context. I would suggest that phonetics is the field from which we could learn a great deal, frequency analysis of nervous-output being very similar to frequency analysis of a phoneme.

Zottermann : I believe that olfactive discrimination is fundamentally a property of receptors.

Schneider : No, I must disagree, since the number of receptors is less than the number of distinguishable sensations. These receptors are more or less specialized, and each cell is a little computer, sending (or not sending) just one frequency modulation to the brain. We must not however forget the smaller number of neurons in the bulb (several thousands) compared with some five millions receptors. There is clearly something happening between these levels, and that is why these recordings from the region of the bulb are very interesting, but for the moment remain a mathematical game which does not yet help us.

Kovats : a) What is the correlation between certain properties of the substance and the olfactive sensation that they induce?

b) How do these receptors function? that is to say how are these "certain properties" transmitted to the receptors? There must be an interaction between the latter and the odorous molecule which cannot be different from that of the type of binary mixture occurring in gas chromatography, as Dr. Dravnieks has suggested.

Eykelboom : Could an animal react differently to a certain flavor or perfume chemical from us?

Mac Leod : An animal probably recognizes the differences as well as us, but they do not interest it. The means of reception (possibly by proteins) are probably the same in both animal and man.

Kalmus : The same object is designated in different languages spoken by man by words which are physically different but having the same meaning; in the same way, the osmoreceptor system of each individual could elaborate responses that are specific to him, but which would be impossible to represent physically because their deep meaning escapes us. What we are recording is not a mathematical synthesis of what receptors are doing, but a complex elaborated by many stages of our central nervous system, so that the property we perceive is not a property of the stimulus at all, but of this elaboration, finally producing what we call the sensation. The same substance can thus cause sensations that are apparently identical, or apparently different, according to the animal species considered.

Zottermann : I agree. For example, for the dog, saccharine is not sweet but bitter, although the dog has sweet receptors. Everything depends on the way the brain interprets information transmitted by the receptors.

Fischer : I also agree with Prof. Kalmus. I think we are mixing up two different levels. What is conveyed at the receptor level might be called "information", and what is interpreted by the organism as a self-referential system is "meaning". For the dog, the information "saccharine" signifies bitter, for us it means sweet.

Hughes :   One receptor alone will not tell us about descrimination, so we have two possible approaches.

1) We may have many micro electrodes in single elements, simultaneously following activity from many single cells, or

2) As we ourselves work, on account of the very great difficulty of method 1, a
macro-electrode approach, in which we are looking at the response of thousands of cells. The information is at the receptor level, but you must look at
many receptors at the same time to understand discrimination.

SPECIALIZED ODOR RECEPTORS OF INSECTS

Dietrich Schneider
Max-Planck-Institut für Verhaltensphysiologie,
8131 Seewiesen, Germany.

Motto from my guest book :

> There was a young man from Seewiesen
> Who sought to discover the reason
> Why the Duft of a butterfly
> Unnoticeably flutters by.
> Aber es war unmöglich zu bewiesen.

> Nicholas Wade, June 26, 1970.
> (a young English colleague – and not a
> chemo-receptionist – who had a two year
> stopover at Seewiesen on his trip from
> Australia to Scotland).

Reception of chemical signals might very well be the earliest sensory mechanism which was ever developed in living organisms. Signals of a chemical nature do, of course, come from the living and not living surroundings of the organism. Already the relatively primitive bacteria and protozoans can readily distinguish between different chemicals as they show by their feeding and avoidance behavior. Furthermore, intraspecific recognition by chemical signals was probably an essential factor in general, and in the development of sexuality in particular. But also the inter-specific prey-, and predator-detection by chemoreception must go back to the earliest times of life.

While it appears that chemical **information transfer** is the more general mechanism which includes all kinds of chemoreception, **communication by chemical signals** is the mechanism restric-ted to behavior where the sender is "interested" in conveying information to the receiver, for exam-ple, the gynogamete emits a compound to attract an androgamete.

With the insects, which are the objects of our own study, we find a wealth of phenomena of the types described so far. Not only are animals of this class able to orientate in relation to the chemical composition of their surroundings, but they also exhibit an extremely sophisticated array of inter- and intraspecific signals.

Communication as defined extends from the insects in the more general situations where flowering plants lure their insect pollinators by scents, to highly specialized mechanisms of intra-specific chemical attraction. It becomes more and more clear that scented flowering plants develo-ped phylogenetically in an intimate dependency of the insects and vice versa. Mostly, the insect pollinators are rewarded by nectar or pollen for their services.

Intraspecific chemical communication is performed by chemical signal compounds often produced by only one sex. Messenger compounds of this kind are now called **pheromones** and are of different biological types : sexual pheromones (attractants, stimulants, arrestants, aphrodisiacs), alarm pheromones, assembly pheromones, trail pheromones, and marker pheromones. While it is obvious from our knowledge of insect behavior that the majority of the insects do have female sexual attractants (probably all the moths), it appears that the other pheromone types are not so evenly distributed. The greatest variety of pheromones is found among the social hymenoptera and possibly the termites. Among the lepidoptera, species-specificity of the attractants and arrestants is normally restricted to groups of related genera (Schneider 1962, Priesner 1969, Schneider and Seibt 1969), but there are exceptions, where two species, which are morphologically and ecologically very similar, do have attractants which are the stereo-isomers of one compound (Roelofs and Comeau 1969). All the physiological evidence available so far indicates that the releaser pheromones of insects – which are the signal compounds received by a sensory mechanism – activate specialized receptor cells of very high sensitivity.

The biologically best known pheromone system is that of the honey bee (see Butler 1969). In this species the female sex attractant (queen substance = 9-oxodecenoic acid) is not only functional to lure the males (drones) to the queen during the nuptial flight, but also to lure the worker bees to follow the young queen during swarming. Another pheromone which is also produced in the queen's mandibular gland is 9-hydroxydecenoic acid, which plays a related, but specific role during swarming. This compound also inhibits the workers from raising young queens in the hive. A mixture of compounds, with geraniol as the main component (Boch and Shearer 1962), is the pheromone of the Nasanoff-gland of the bee, which regulates a number of activities. There are a few more bee pheromones among which alarm substances are the most interesting. If, for instance, a worker bee is molested near the hive, she emits a chemical alarm, which makes other workers attack and sting all moving, dark objects. The known alarm pheromones are isoamyl acetate and 2-heptanone (see Boch et al. 1969) from two different odor glands. A few more bee pheromones are markers for rich and poor food sources, or the recently detected bee-footprint odor (see Butler 1969). Pheromone research is a young discipline which is based on old and new behavioral evidence. Chemical analysis, as well as physiological studies of the receptor systems, are rather recent advances.

The following examples are from published and unpublished work which was done in our laboratory by different investigators.

SEXUAL ATTRACTANT RECEPTION IN THE SILKWORM MOTH BOMBYX

The *Bombyx* attractant is produced in female abdominal glands, the sacculi laterales (Steinbrecht 1964 a and b), and was the first pheromone to be identified chemically. It is a C_{16}-alcohol : hexadeca-10-*trans*, 12-*cis,* dien-1-ol (Butenandt et al. 1959, 1961). This substance, which was called **bombykol**,was synthesized together with its geometrical isomers (Butenandt et al. 1959, 1962). Recently it has also been prepared in tritiated form (Kasang 1968). Thanks to this chemical work, detailed physiological studies could be performed. These investigations started fifteen years ago (Schneider and Hecker 1956) and have been continued since then. Because the bombykol work was recently summarized by several authors (Schneider 1969, Kaissling and Priesner 1970, Schneider 1970 a and b) it will suffice here to outline the major results.

The **sensory system** responsible for the reception of bombykol is located on the antenna of only the male insect. The specialized odor receptors are sense cells in connection with long hair sensilla. The structural organization of one such sensillum trichodeum is typical for the sequence of events which lead to the excitation of the receptor cell. Bombykol receptor hairs are positioned on the combed *Bombyx* antenna in such a way as to filter the odor molecules most effectively out of the air (Fig. 1; Adam and Delbrück 1968). Bombykol molecules are readily adsorbed on the air surface. From here they are thought to travel by means of surface diffusion to openings which break through the hair cuticle (Fig. 2). These pores are continuous with a system of pore canals

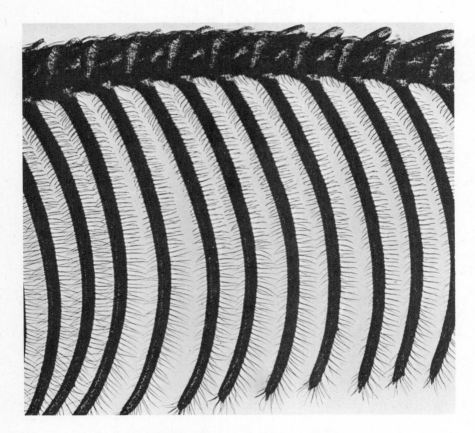

Fig. 1. *Bombyx mori* ♂. Part of a male antenna showing the arrangement of
olfactory sense hairs on the antennal branches. The average length of
the sense hairs is 100 μ. (Steinbrecht 1971).

which reach into the fluid-filled hair-lumen, where they end in proximity (not yet precisely defined) to the receptive dendrites of the sensory cell. Should this fluid (the "sensillum liquor", Ernst 1969) be interspaced between the canals and the receptor-cell membrane, one would have to assume volume diffusion of the molecules to take place before they reach their destination. This sequence of events is verified by direct of indicative evidence of different kinds, namely morphological studies, ^{3}H-bombykol adsorption measurements, electrophysiological recordings from whole antennae and single cells, and behavior-threshold measurements (Kaissling and Priesner 1970).

The **behavior threshold** of the male moth is reached when each antenna adsorbs about **500 bombykol molecules per 0.86 sec.,** which is the mean reaction time of the male moth.

How many molecules are now under these circumstances "caught" by the hairs and eventually available for the receptor process? There are 16,000 hair sensilla per antenna, with 25,000 sensory cells of the type involved in this reaction. All these cells together receive 300 molecular

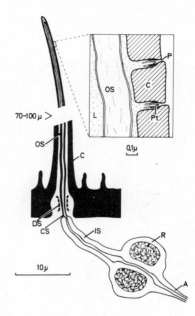

Fig. 2. *Bombyx mori* ♂. Schematic diagram of a sensillum trichodeum in longitudinal section. The sense hair is innervated by two bipolar receptor cells (R), the receptor processes of which show three morphologically different segments : the inner segment (IS), the ciliary segment (CS), and the outer segment (OS). Inset : The pore-tubule systems in the cuticular wall (C) of the sense hair probably functions as stimulus-transfer system for odor molecules. A = axon, DS = dendrite sheath, L = "Sensillenliquor", P = pore, Pt = pore tubule. (Courtesy of Dr. R.A. Steinbrecht).

hits of the 500 molecules adsorbed (Fig. 3). These **300 molecules elicit about 200 nerve impulses** more than the spontaneous firing rate. This now suggests that the cells actually respond to single molecule hits and do not need double hits to produce meaningful signals for the brain. Such a deduction is supported by the following facts and calculations :

1. If one assumes that the molecule-receptor encounter is a random process, Poisson statistics are the adequate checking procedure. One can now say that the three hundred molecules available are contacting about the same number of cells with a single hit each and only two cells have a chance to receive a double hit. The justification for the use of Poisson statistics comes from single cell recordings (Fig. 4). A plot of single receptor-cell impulses as a function of stimulus strength, shows that these curves are identical with single- and double-hit Poisson curves (Fig. 5). Because of this and the low number of molecules which are totally available, the hit process is definitely random. Thus it appears from these observations that single molecule-impacts are eliciting single impulses, and two impacts two impulses. With higher stimulus intensities, the relationship is more complex.

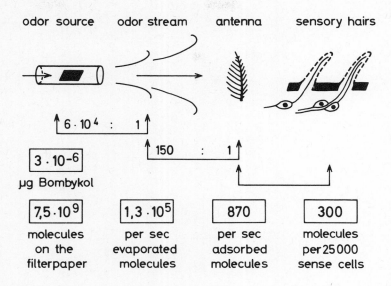

Fig. 3. Determination of molecule numbers of tritiated bombykol adsorbed on the *Bombyx* antenna. 3×10^{-6} μg is the load of the odor source eliciting a significant behavioral response. The number of 300 molecules is calculated for the average reaction time of the animals (0.86 sec). (From Kaissling and Priesner 1970).

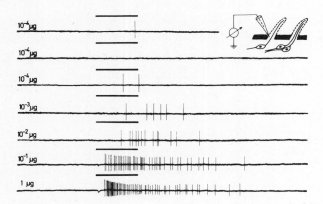

Fig. 4. Impulse response of a bombykol receptor of the male silkworm moth. RC-recording (glass capillaries) from a sensillum trichodeum (scheme right above). With 10^{-3} μg bombykol on the odor source, approximately 7 molecules per second stimulus duration are available for one receptor cell. (From Kaissling and Priesner 1970).

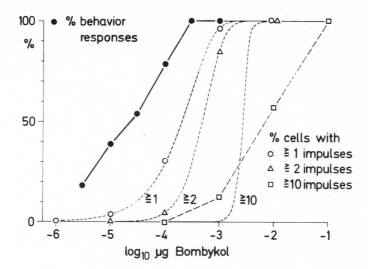

Fig. 5. Evaluation of several hundred impulse recordings at different loads of the
odor source (see Fig. 4). The percentage of measurements with $\geqslant$ 1 and
$\geqslant$ 2 impulses fits with theoretical Poisson curves (hatched lines). Only
a small percentage of cells fire at the behavioral threshold (left curve). The
spontaneous activity of the cells was substracted (from Kaissling and Priesner,
1970).

2. The electrophysiologically recorded spontaneous "noise" of the bombykol receptor cells of
one antenna is of the order of 1,600 impulses per antenna. Since at the behavior threshold
only two of the cells (0·005 %) received double hits, the meaningful brain information can
not possibly be the result of double hits (and paired impulse) responses. The signal to noise
ratio is for the single hits 200/1,600. Information theory requires the minimum signal to be

$$3 \times \sigma = 3 \times \sqrt{1600} = 120.$$

This value of approximately 200 impulses per second is enough to secure the safe transfer of
the message : "bombykol is in the air".

These experiments show the bombykol receptor and the whole antennal receptor system
to be as sensitive as possible, which means that we are dealing with highly triggered systems. In addition, the catching capacity of the complex antennal geometry is well suited to serve this biologically highly adapted receptor system.

The most puzzling fact in this series of mechanisms is still the obviously very efficient molecule transport system which involves the diffusion of the bombykol particles from the sense hair
surface to the receptor site (the "acceptors") on the dendrite. But the time calculated to be maxi-
mally necessary for this process is much shorter (150 msec) than the reaction time of the impulses
(500 msec) (compare : Adam and Delbrück 1968, with Kaissling and Priesner 1970).

How specific is the bombykol receptor system? Earlier and more recent recordings gave
some insight into this problem (Schneider et al. 1967; Priesner 1969). Some of the alcohols, which
are related to bombykol, need to be about 10^6 times more highly concentrated than bombykol to

reach the same response level (Fig. 6). Even for the synthetic *trans, trans* bombykol-isomer, the threshold is elevated 1000-fold. So far, we have no explanation for this phenomenon which is, of course, an expression of our ignorance of the acceptor material of the receptor cells. Kaissling (1969) has considered the kinetics of the receptor system, which are not necessarily the kinetics of the acceptor-bombykol interaction. These calculations are based on the slow, summarized receptor potential, the electroantennogram (EAG) and lead to formulas which could either indicate that adsorptive or enzymatic processes are involved.

Hexadeca-10-*trans*,12-*cis*-dien-1-ol	0.001
Hexadeca-10-*cis*,12-*trans*-dien-1-ol	0.1
Hexadeca-10-*cis*,12-*cis*-dien-1-ol	1
Heptadeca-10-*cis*,12-*cis*-dien-1-ol	10
Hexadeca-10-*trans*,12-*trans*-dien-1-ol	1
Hexadec-10-*trans*-en-12-yn-1-ol	100
Hexadec-10-*trans*-en-1-ol	1
Hexadec-9-*cis*-en-1-ol	100
Octadec-9-*cis*-en-1-ol	1000
Octadeca-9-*cis*-12-*cis*-dien-1-ol	1000
Octadec-9-*trans*-en-1-ol	> 1000
Octadec-11-*trans*-en-1-ol	> 1000

Fig. 6. *Bombyx mori* ♂. Sex pheromone receptor thresholds to 12 primary unsaturated alcohols, as obtained by EAG and single cell recordings. The threshold values are stated in μg of stimulus source concentration. They were determined by increasing concentration steps logarithmically from 10^{-6} μg to 10^{-3} μg. (From Priesner 1969).

Our attempts to decide about the fate of the bombykol molecule in the process of transduction are not very far advanced. Kasang (see this symposium) has found a metabolic degradation product of bombykol on the antennae of both sexes and other body parts of *Bombyx* which was quite effective. On account of a slow time constant this could not result from an interaction of bombykol with the acceptor. This process might, however, play a very important role in a number of respects, and may, for instance, cause avoidance of self-stimulation of males by odor desorbed from their own body as a secondary odor source.

GRASS-ODOR RECEPTOR OF THE LOCUST

Migratory locusts (*Locusta migratoria*) exhibit positive anemotactic behavior when an air current contains the smell of green plants (Moorhouse, personal communication). In search for the receptor which may initiate such motor response, Boeckh (1967 a) detected that receptor cells of antennal pit-sensilla *(s. coeloconica)* respond to green plants, aldehydes (which are ubiquitous in green leaves) and fatty acids. He determined the receptor threshold, and tested the stimulatory effectiveness of some 80 compounds.

This type of sensillum, which is for a number of reasons very well suited for an electrophysiolocal analysis of the odor response spectrum of single receptor cells, has again been studied with improved techniques. Kafka (1970 and this symposium) selected hundreds of compounds according to their physical-chemical properties and tested them with this locust receptor. The final results of this thorough study are that compounds which are effective stimulants for this receptor cell need to fit two requirements (i) The polar molecule has to have a chain-like supporting ("substrate") part which is probably important for weak bonding effects with the supposed acceptor. This part of the molecule may have the task of facilitating approach and attachment to the acceptor. (ii) The odor molecule has to have an "active structure" which is thought to be decisive for the initation of the transduction and may temporarily go through a stronger, but not yet co-valent, bonding phase to a special site on this acceptor. This cell is up till now the most extensively studied odor receptor cell we know. The results of Kafka's work allow us to examine critically most of the so called odor theories, because here we can really judge the properties of the odor molecules which are stimulants for one given type of receptor cell. So far, all theoretical considerations could only be based on the comparison of a wide variety of effective stimulants to which the human nose responds, or a whole animal reacts.

The structure of the grass receptor-organ of the locust is rather different from the type of organ described for *Bombyx*, and is described by Kafka in the next paper of this symposium. The cuticular cone of the locust's coeloconic sensillum – which is homologous to the bombycol receptor hair – has a fluted wall which is penetrated by canals filled with electron-dense material. These are probably the channels which lead the odor molecule to the dendrite (Steinbrecht 1969). The molecule-catching properties of the pit-sensillum are obviously of a different quantitative dimension as compared with the *Bombyx* sensory hairs (see also Kaissling 1971). With such a small cone, adsorptive chances for odorants are markedly reduced. This, of course, does not mean that the receptor cell's threshold is not as well developed as in *Bombyx*. Kafka (loc. cit.) could show that very probably, single molecules are sufficient to elicit responses in this cell.

Apparently, a different selective pressure was effectively "shaping" this sensillum in relation to the biological needs of this insect. The whole antenna of the locust is a slender rod, without branches, looking much more like a "feeler" than the *Bombyx* antenna. The filter capacity of such an antenna is of necessity comparatively poor, which again reduces the probability of this organ catching the odor molecules (Kafka 1970; see also Kaissling 1971).

The result of selective adaptation in phylogeny is always a compromise. *Locusta* must be able to "afford" (or find it useful to have?) a poorer molecule filter system on the one hand, in order to have on the other hand a slender antenna. A rod shaped antenna may in addition be an unescapable design for the hemimetabolic insect, where the instars have antennae similar to the imago. The total number of olfactory cells on the antenna of the locust is small compared to a feather antenna of a moth, but also small compared to the honeybee antenna. The bee antenna is superficially very similar to the locust antenna and very probably has a similar molecule-catching capacity. However, the bee antenna is much more densely packed with olfactory receptor cells which even outnumber the cells in the moth antennae. Drone antennae are reported to have up to 500,000 olfactory cells with their pore plate sensilla (for references see Lacher 1964) while in the saturniids and *Bombyx*, we found only less than 100,000 olfactory cells per antenna (Schneider and Kaissling 1957, Boeckh, Kaissling and Schneider 1960, Steinbrecht 1971).

RECEPTORS FOR THE SEXUAL ARRESTANT (APHRODISIAC) OF A BUTTERFLY

Butterflies of the family Danaidae are well known as mimicry models. Recently, behavioral, biochemical and physiological work concentrated on courtship behavior of these insects, which is not only controlled by visual signs but also by a very sophisticated display of a male pheromone, which was called an aphrodisiac (Pliske and Eisner 1969).

In this case, the pheromone-receptor analysis is not nearly as advanced as in the cases just mentioned, but nevertheless is a good example of a biologically interesting transmitter-receiver system.

Fig. 7. *Crastia amymone* (Godt.) (Lepidoptera, Danaidae) from Hongkong. Fully expanded abdominal hairpencils of a male butterfly. The hairs are hollow and connected to a gland which contains the pheromone. The animal was kindly supplied by Dr. Daisy Leong.

The male Florida queen-butterfly *(Danaus gilippus)* is equipped with a pair of abdominal hair brushes, which can be actively expanded (Fig. 7). The long glandular hairs of the organ are designed to distribute the pheromone during the courtship flight, when the male tries to seduce the female by use of his "perfume". When hovering in front of the female, the male expands his odor organs and virtually strews pheromone-containing dust onto the female's antennae. The female consequently settles on foliage and the pair mate (Brower et al. 1965, Pliske and Eisner 1969).

In a chemical analysis of the queen hair-pencils, two major components were found : (I) a crystalline ketone (2,3-dihydro-7-methyl-1H-pyrrolizin-1-one); (II) a viscous terpenoid alcohol (3,7-dimethyldeca-2-*trans*,6-*trans*-dien-1,10 diol) (see Meinwald et al. 1968, 1969).

Behavioral and electrophysiological tests showed the ketone to be the olfactorially effective compound, while the diol seems to work as a fixative, somewhat in the same way as in perfume technology (Pliske and Eisner 1969, Schneider and Seibt 1969). The fixative, but not the ketone, can experimentally be replaced by other sticky compounds, as for instance mineral oil, without any interference with the result.

The dust particles are products of the gland-hair surface; they are of cuticular origin and contain substances I and II (Fig. 8). They only stick to the female's antenna if the fixative, working like a glue, is present, because dust without the diol (or a substitute) is easily lost,and males with dust which contains only the ketone are only weakly effective in courtship (Pliske and Eisner 1969).

In electrophysiological experiments, we found that the ketone-gas is eliciting electroantennograms in the antennae of both sexes. Partial covering of the antenna with wax corroborated

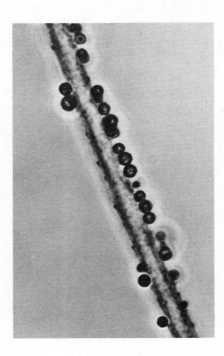

Fig. 8. *Danaus gilippus berenice* (Cramer) (Lepidoptera, Danaidae). Florida
Queen butterfly. Single hair of the male abdominal hairpencil with
cuticular "dust" particles. The hair and the dust are impregnated with
the pheromone and a fixative. Hair diameter : approx. 20 μ. The Queen-
butterflies were kindly supplied by Drs. Th. Eisner and Th. Pliske.

the statement of J. Myers (1968) that a special type of sensillum basiconicum, which is found widely distributed on the club-shaped antenna, is the sensory organ responsible for this reaction (Myers
and Brower 1969).

Single-cell recordings are technically very difficult in these butterflies, but the EAGs permit
us to say that both sexes do have the ketone receptors. The diol elicits only poorly reproduceable
and weak EAGs, and does not play a major – if any – role as an odorant. In order to test the effect
of dust with or without any of the compounds I and/or II, we carried out the following experiments.
The antenna of the *Danaus* butterfly was mounted between electrodes and a glass cartridge was
loaded with dust and positioned in front of the antenna. With a puff of air, the dust was blown onto
the antenna and the response recorded. With "clean" dust, only a mechanical artifact was observed.
Dust which was only impregnated with the ketone gave a full sized EAG which did not last much
longer thant the EAG elicited by ketone gas. If, in addition to the ketone, the diol was added to the
charge, the EAG lasted much longer and many more dust particles adhered to the antenna than in
the former experiment. If instead of the diol, mineral oil was used as a fixative, the same effect was
observed. On the other hand, EAGs elicited by dust which contained only the fixative, gave responses which did not differ from the control (Schneider and Seibt 1969). While qualitative single cell
spectra of *Danaids* are not available, EAG studies of antennae stimulated with the pheromone and
some related compounds gave a first insight into this problem. A number of cyclic nitrogen-containing substances were to a varying degree also effective stimulants for these preparations (Fig. 9)

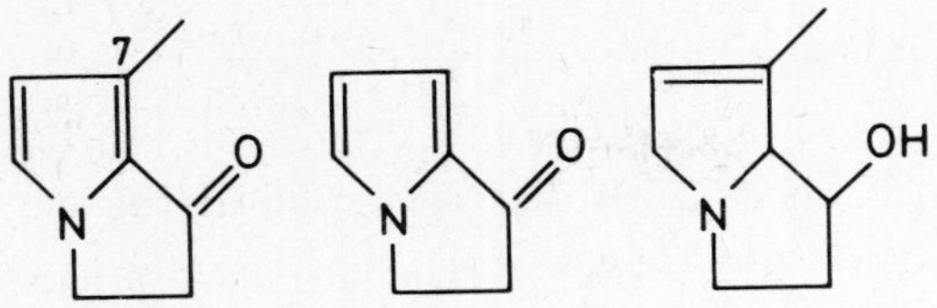

2,3-Dihydro-7-methyl-1H-pyrrolizin-1-one

Fig. 9. Substances tested in the electro-antennogram (EAG) with Danaid-butterflies (see text). The compound on the left side (see the written name) is the ketone which was found to be the natural male aphrodisiac (arrestant). It was fully effective as an odorant on all 7 danaid species tested so far. The compound in the middle was equally effective, while the one on the right side had no effect. The synthetic compounds were kindly supplied by Dr. J. Meinwald.

Finally, it appeared that all the Danaids of several genera so far tested, seem to have the same ketone as male aphrodisiac (substance I). Chemical evidence for this is available for *Danaus gilippus* (Florida) and *Lycorea ceres* (Trinidad). Full EAG reactions in cross tests with the ketone and glands of *Danaus gilippus, Lycorea ceres, Danaus chrysippus* (Africa), *Tirumala (Danaus) limniace* (Hongkong), *Radena similis* (Hongkong), *Salpinx (Euploea) midamus* (Hongkong) and *Crastia (Euploea) amymone* (Hongkong) have been observed. This indicates that over the whole family Danaidae the same male pheromone – or a closely related and equally effective – compound is produced. Final proof for this claim can, of course, only come from chemical work. The North-American *Danaus plexippus* (the Monarch) has olfactorily ineffective, small hairpencils which neither contain the ketone nor the fixative (Meinwald et al. 1969/70), but the antenna responds with the full EAG to the ketone or the glands of all the other Danaids. Here we may be dealing with a phylogenetically interesting case, where the pheromone production was abandoned (or changed in a way not yet understood) in the male sex, while the receptor is still "surviving" as a "rudiment". The courtship of the Monarch corresponds to this hypothesis. The pair mates high in the air after an obviously optically controlled very rapid pursuit (personal communication by T. Pliske).

Extensive comparative studies with female attractant pheromones of hundreds of different moth species also led to the conclusion that the releaser pheromones are by no means species specific. Priesner (1969) assumes that among more than one thousand species in the family Saturniidae, only some forty different female attractants may be present. In these cases, we also still lack the chemical proof of this claim, but the physiological experiments are clearly in favor of such a deduction.

CONCLUDING CONSIDERATIONS

For a number of reasons, olfaction is a challenging biological phenomenon. One aspect of this has been outlined in the introductory chapter of this paper; it is the phylogenetically very old mechanism of recognition and differentiation by means of molecular patterns with such a high level of sophistication in the pheromone system.

The interaction of a given signal-molecule with an acceptor in the cell membrane is obviously a problem which is analogous – if not identical – to well-known processes like, for instance, enzyme-substrate interaction. The process of chemoreception, however, is more than simply a bimolecular encounter, because it also involves integrative functions on the intra- and intercellular level. Even a bacterium, a flagellate, or a ciliate protozoan is able to make differential use of two or more chemical signals, which means that the effector system of this cell is steered by more than one input channel.

The simplest design for such a system of a unicellular organism would be that the acceptors in the cell membrane are not uniform but of as many types as there are chemical signals the cell is differently responding to. The acceptors are here assumed to be macromolecular systems which regulate the ion-permeability of the cell membrane and the effectors to be cilia which beat differently according to the ion-controlled membrane polarization. In this way, the protist cell would be able to activate its respective effectors in different ways. The same result could also be reached, if one and the same acceptor would be able to respond in a two-way fashion to either a molecule of type A and/or type B. The acceptor is in this case thought to be one single gate, which may now be able to let different ions penetrate, depending upon the type of stimulus.

With the higher forms of animals we find well developed chemoreceptor systems where specialized cells serve as the receptor elements. The whole organisms – as we see in the insects and vertebrates – are now able to differentiate among a rather large number of chemicals as is typical in olfaction. One may divide the classical **"odor problem"** into **three sub-problems** :

1. The interaction of the odor molecule with the acceptor in the cell membrane.

2. The membrane permeability changes which are induced by the first process and eventually lead to the elicitation of a receptor potential and the following nerve impulses (involving the stimulus-response transduction).

3. The integrative function of the nerve centers which deal with the information flux from the peripheral olfactory cells.

Sub-problem (1) is familiar to us from the example of the single cells, which may, however, be much more complicated systems than outlined here. The problem is clearly one of physico-chemical nature.

Sub-problem (2) is also of this nature but involves in addition a higher organization because a receptor cell integrates the signals of its acceptors spatially and temporally. It is known from the insect olfactory-cells, that they have in some cases more than one type of acceptor in the dendritic membrane. The result of the activation of these acceptors may either be an excited state following the interaction with any of the effective stimulants, or in other cases an excited or an inhibited state, again depending upon the chemical stimulus (see Boeckh 1962, 1967 b, et al. 1965, Schneider et al. 1964, Kafka 1970). The cell is , however, a unit and the final impulse message is the result of its integrative function.

Sub-problem (3) is the least understood so far. We are fairly ignorant of the functional properties of the olfactory nerve centers. Morphological studies show us that the number of secondary olfactory nerve cells in insects and vertebrates is much smaller than the number of receptor cells. Does this mean that each secondary cell receives only information from receptor cells which have identical reactivity? In vertebrates, it does not look so simple. Unfortunately, vertebrate physiologists know something of the reaction spectra of the secondary (mitral) cells, but nearly nothing of the peripheral cells. On the other hand, insect physiologists have ample information about receptor cell spectra but no knowledge of the responses of the secondary cells. It is obvious that sub-problem (3) is not a problem of molecular biology, but one of the processing of nervous information. For a full solution of the odor problem, we **need to decipher the "odor code" on all levels** : the **molecular,** the **cellular,** and the **central.** With the insects, we have made some progress on level (1) and (2). With suitable insect species, where behavioral studies gave us information on the power of odor-discrimination during the animal's behavior, electrophysiological recordings from central cells might in the future help us to an understanding of level (3). Central nervous system recordings from insects are possible as I have seen myself and as was recently demonstrated by Yamada (1968). But such studies can of course only lead to a deeper understanding if they are done with animals where one knows their olfactory behavior and function on level (1) and (2).

The **specialized receptor cells** – which are the topic of this paper, are certainly the organs allowing insects to respond to chemically uniform signals of special biological importance. But are they also the means by which the insect differentiates among a multitude of odor signals? To do this, the animal would need a very large number of types of specialized odor receptors. This number could be drastically reduced if the cell's reaction spectra to different chemicals overlap. In this case one could imagine that a relatively simple system can decode a complicated pheripheral pattern of messages. But did nature use these ideas? We found large numbers of differently reacting odor receptor cells in some insects and gave them the "working title" of "odor generalists" (Schneider et al. 1964). However these studies must be extended and, as we have seen, be related to the levels (1) and (2) .

It appears that we are making some progress in our understanding of the membrane and receptor cell mechanisms, but still badly lack information on the functional integration of these processes. Hopefully, the insect is a good model system to make further progress but the future will show whether we are right.

REFERENCES

Adam, G. and Delbrück, M. 1968. Reduction of dimensionality in biological diffusion processes. In "Structural chemistry and molecular biology", A. Rich and N. Davidson, Eds., Freeman, San Francisco and London, p. 198-215.

Boch, R. and Shearer, D.A. 1962. Identification of geraniol as the active component in the Nassanoff pheromone of the honey bee. *Nature, London 194,* 704-706.

Boch, R., Shearer, D.A. and Petrasovits, A. 1969. Efficacities of two alarm substances of the honey bee. *J. Insect Physiol. 16,* 17-24.

Boeckh, J. 1962. Elektrophysiologische Untersuchungen an einzelnen Geruchsrezeptoren auf den Antennen des Totengräbers (*Necrophorus,* Coleoptera). *Z. vergl. Physiol. 46,* 212-248.

Boeckh, J. 1967 a . Reaktionsschwelle, Arbeitsbereich und Spezifität eines Geruchsrezeptors auf der Heuschreckenantenne. *Z. vergl. Physiol. 55,* 378-406.

Boeckh, J. 1967 b. Inhibition and excitation of single insect olfactory receptors and their role as a primary sensory code, II. Int. Symp. Olfaction and Taste, T. Hayashi, Ed., Oxford, Pergamon Press 721-735.

Boeckh, J., Kaissling, K.-E. and Schneider, D. 1960. Sensillen und Bau der Antennengeissel von *Telea polyphemus* (Vergleiche mit weiteren Saturniden : *Antheraea, Platysamia* und *Philosamia*). *Zool. Jb. Anat. 78,* 559-584.

Boeckh, J., Kaissling, K.-E. and Schneider, D. 1965. Insect olfactory receptors. Cold Spring Harb. Symp. Quant. Biol. 30, 263-280.

Brower, L.P., van Zandt-Brower, J. and Cranston, F.P. 1965. Courtship behavior of the Queen butterfly, *Danaus gilippus berenice* (Cramer). *Zoologica (New York) 50,* 1-39.

Butenandt, A., Beckmann, R., Stamm, D. and Hecker, E. 1959. Ueber den Sexual-Lockstoff des Seidenspinners *Bombyx mori.* Reindarstellung und Konstitution. *Z. Naturforsch. 14 b,* 283-284.

Butenandt, A., Hecker, E., Hopp, M. and Koch, W. 1962. Die Synthese des Bombykols und der *cis-trans*-Isomeren Hexadecadien-(10.12)-ole-(1). *Liebigs Ann. Chem. 658,* 39-64.

Butler, C.G. 1969. Some pheromones controlling honeybee behaviour. Proc. VI Congr. IUSSI, Bern, 19-32.

Ernst, K.-D. 1969. Die Feinstruktur von Riechsensillen auf der Antenne des Aaskäfers *Necrophorus* (Coleoptera). *Z. Zellforsch. 94,* 72-102.

Kafka, W.A. 1970. Specificity of odor molecule interaction in single cells. This symposium.

Kafka, W.A. 1970. Analyse der molekularen Wechselwirkung bei der Erregung einzelner Riechzellen. *Z. vergl. Physiol.* in press.

Kaissling, K.-E. 1969. Kinetics of olfactory receptor potentials. III. Int. Symp. Olfaction and Taste 1968, C. Pfaffmann, Ed., Rockefeller University Press, New York, 52-70.

Kaissling, K.-E. Insect olfaction. Handbook of sensory physiology, Springer Verlag, in press.

Kaissling, K.-E. and Priesner, E. 1970. Die Riechschwelle des Seidenspinners. *Naturwissenschaften 57,* 23-28.

Kasang, G. 1968. Tritium-Markierung des Sexuallockstoffes Bombykol. *Zeitschr. Naturforsch. 23b,* 1331-1335.

Kasang, G. 1970. Bombykol reception and metabolism on the antennae of the silkmoth *Bombyx mori.* This symposium.

Lacher, V. 1964. Elektrophysiologische Untersuchungen an einzelnen Rezeptoren für Geruch, Kohlendioxyd, Luftfeuchtigkeit und Temperatur auf den Antennen der Arbeitsbiene und der Drohne *(Apis mellifica* L.). *Z. vergl. Physiol. 48,* 587-623.

Meinwald, J., Chalmers, A.M., Pliske, T.E. and Eisner, T. 1968. III. Pheromones Identification of *trans, trans*-10-hydroxy-3,7-dimethyl-2,6-decadienoic acid as a major component in "hair-pencil" secretion of the male monarch butterfly. *Tetrahedron Letters 1968,* 4893.

Meinwald, J., Meinwald, Y.C. and Mazzocchi, P.H. 1969. Sex Pheromone of the Queen Butterfly: Chemistry. *Science 164,* 1174-1175.

Myers, J. 1968. The structure of the antennae of the Florida Queen Butterfly, *Danaus gilippus berenice* (Cramer). *J. Morphol. 125*, 315-328.

Myers, J. and Brower, L.P. 1969. A behavioral analysis of the courtship pheromone receptors of the queen butterfly, *Danaus gilippus berenice. J. Insect Physiol. 15*, 2117-2130.

Pliske, T.E. and Eisner, T. 1969. Sex pheromone of the queen butterfly : Biology. *Science 164*, 1170-1172.

Priesner, E. 1969. A new approach to insect pheromone specificity. III. Int. Symp. Olfaction and Taste 1968, C. Pfaffmann, Ed., Rockefeller Univ. Press, New York, 1969, 235-240.

Roelofs, W.L. and Comeau, A. 1969. Sex pheromone specificity : Taxonomic and evolutionary aspects in lepidoptera. *Science 165*, 398-400.

Schneider, D. 1962. Electrophysiological investigation on the olfactory specificity of sexual attracting substances in different species of moths. *J. Ins. Physiol. 8*, 15-30.

Schneider, D. 1969. Insect olfaction : Deciphering system for chemical messages. *Science, 163*, 1031-1037.

Schneider, D. 1970 a. Insect communication by means of pheromone molecules. In press in : "Theoretical Physics and Biology". Proc. Second Internat. Conference. Paris/Versailles (1969), North-Holland, Amsterdam and Interscience (Wiley) New York.

Schneider, D. 1970 b. Olfactory receptors for the sexual attractant (Bombykol) of the silkmoth. In press in : "The Neurosciences", Second Study Program. F. O. Schmitt, Edit. in Chief, Rockefeller Univ. Press.

Schneider, D., Block, B.C., Boeckh, J. and Priesner, E. 1967. Die Reaktion der männlichen Seidenspinner auf Bombykol und seine Isomeren : Elektroantennogramm und Verhalten. *Z.vergl. Physiol. 54*, 192-209.

Schneider, D. and Hecker, E. 1956. Zur Elektrophysiologie der Antenne des Seidenspinners *Bombyx mori* bei Reizung mit angereicherten Extrakten des Sexuallockstoffes. *Z. Naturforschg. 11b*, 121-124.

Schneider D. and Kaissling, K.-E. 1957. Der Bau der Antenne des Seidenspinners *Bombyx mori* L. II. Sensillen, cuticulare Bildungen und innerer Bau. *Zool. Jb. Anat., 76*, 223-250.

Schneider, D., Lacher, V. and Kaissling, K.-E. 1964. Die Reaktionsweise und das Reaktionsspektrum von Riechzellen bei *Antheraea pernyi* (Lepidoptera, Saturniidae). *Z. vergl. Physiol. 48*, 632-662.

Schneider, D. and Seibt, U. 1969. Sex Pheromone of the Queen Butterfly : Electroantennogram responses. *Science 164*, 1173-1174.

Schneider, D. and Steinbrecht, R.A. 1968. Checklist of Insect Olfactory Sensilla. *Symp. zool. Soc. Lond. 23*, 279-297.

Steinbrecht, R.A. 1964 a. Die Abhängigkeit der Lockwirkung des Sexualduftorgans weiblicher Seidenspinner *(Bombyx mori)* von Alter und Kopulation. *Z. vergl. Physiol. 48*, 341-356.

Steinbrecht, R.A. 1964 b. Feinstruktur und Histochemie der Sexualduftdrüse des Seidenspinners *Bombyx mori* L. *Z. Zellforschung 64*, 227-261.

Steinbrecht, R.A. 1969. Comparative morphology of olfactory receptors. III. Int. Symp. Olfaction and Taste 1968, C. Pfaffmann, editor, Rockefeller University Press, New York 1969, 3-21.

Steinbrecht, R.A. 1971. Zur Morphometrie der Seidenspinnerantenne : Zahl und Verteilung der Riechsensillen. In preparation.

Yamada, M. 1968. Extracellular recording from single neurones in the olfactory centre of the cock- roach. *Nature 217*, 778-779.

SPECIFICITY OF ODOR-MOLECULE INTERACTION IN SINGLE CELLS

Wolf A. Kafka

Max-Planck-Institut für Verhaltensphysiologie

8131 Seewiesen, Germany

We do not know which properties make a molecule odorous, but we know that single olfactory receptor cells react only to distinct substances (Schneider 1957, 1968, 1969; Schneider and Steinbrecht 1968; Boeckh 1962, 1967; Boeckh et al. 1965; Lacher 1964, 1969; Kaissling and Renner 1968). It should therefore be possible to investigate these properties by testing the reaction of single olfactory receptor cells to different compounds. The effectiveness of these substances can then be compared with their physico-chemical properties. This procedure should provide information about how these properties are related to the reactions of the cells.

THE SENSE ORGAN

Experiments of this kind have been done with the olfactory pit organs (sensilla coeloconica) on the antennae of the migratory locust (*Locusta migratoria*) (**Fig. 1**). A little cone-like sense hair is situated in an open pit. The processes of one to three sensory nerve cells run into this hair. They are surrounded by an electron-dense material (Steinbrecht 1969) and a liquid phase, the "Sensillenliquor" (Ernst 1969). A system of pores leads to the outside.

Boeckh (1967) found some of the odorants to which these cells react and studied their effects quantitatively by electro-physiological methods. He found reactions to odors of fresh grass, short-chained fatty acids, aliphatic unsaturated aldehydes, and alcohols. In all, he tested about 80 substances. Because of the technically well suited sensillum and the relatively simple molecular structures of the active substances, these organs seemed to be appropriate systems for further studies, especially for the analysis of the specificity of interaction between receptors and odoriferous molecules. This paper reports on the reaction (nerve-impulse activity) of these receptor cells to nearly 370 compounds having structures varying in a stepwise manner.

STIMULATION

In order to have comparable conditions it was necessary to test the compounds in graded quantities having a known numbers of molecules. This was achieved using a syringe-olfactometer (Kafka 1970 a). The substances were diluted in 10-fold steps with paraffin oil ("Uvasol", fp. 70^0 C). Small glass bottles containing about 0.1 ml of the solution were placed into a plastic syringe. In this practically closed compartment a saturated vapor pressure builds up according to the dilution rate. When advancing the plunger, part of the gas mixture is blown onto the sense organ which has the recording electrode in place. This syringe olfactometer was calibrated by gas injections into a gas chromatograph (Varian Aerograph, Autoprep 705). Adsorption measurements with radioactive

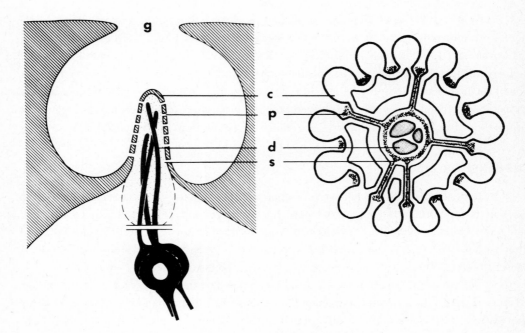

Fig. 1. Pit organ (sens. coeloconicum) on the antennae of the migratory locust in (a) longitudinal and (b) transversal section
(from Steinbrecht 1969); **g**, pit opening; **c**, cuticle; **p**, pore
system; **d**, dendrites; **s**, liquor and electron-dense phase.

labelled substances gave the number of molecules which are adsorbed during one stimulus on an antennal area of the dimension of the pit opening. This number will be called "stimulus quantity", and the impulse frequency generated by the stimulus, "reaction quantity".

MOLECULE REQUIREMENT FOR A CONSTANT RESPONSE

The stimulus quantity necessary to pass the cell threshold level depends on the odorants which are used as the stimulus. The lowest values found to be necessary corresponded to approximately 10 molecules for *trans*-2-hexenal and *trans*-2-hexenoic acid. This implies that cell excitation is brought about by single molecule effects and not by mass effects, like, for instance, those operating in the mechanism of surface tension, viscosity or streaming potentials. Fig. 2 demonstrates the normal stimulus-response increase of a cell when exposed to increasing stimulus quantities. The

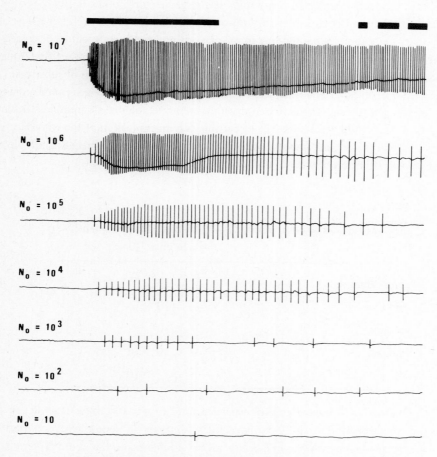

Fig. 2. Typical reaction mode of a single olfactory cell to various
stimulus quantities (N_o = molecules per pit opening) of
trans-2-hexenal. Max. spike amplitude, 4 mV. Black bar:
stimulus time 600 ms, interrupted bar : fresh air (for further
explanations see text).

gradient of this increase (as a function of the stimulus intensity) is independent of the threshold level or, in other words, a nearly equal increase was observed when thresholds for other substances were some factors of ten higher. It might therefore be possible that a given response is always elicited by nearly equal numbers of molecules eventually interacting with an acceptor system, and independent of the stimulus quantity applied to the preparation. That means that differences of the effectiveness of odor molecules are given by different possibilities of interaction with the acceptor system. Specificity is the basic problem of elucidating the reasons for these varying possibilities of interaction and is, as yet, largely unexplained. An acceptor system can be described as consisting of a molecule, or a number of molecules, with which the odor molecules have to interact in order to cause a cell reaction (see below).

SPECIFICITY

Two types of olfactory receptor cells were found in electrophysiological recordings. One type responds to nitrogen-organic, the other to oxygen-organic compounds. Both types of receptor cells could be found in the same olfactory pit organ. Substances which excite one type often inhibit the other. Only the type responding to oxygen-organic compounds will be discussed here.

Fig. 3 shows a small number of the 370 compounds tested. Variations of functional groups are arranged in the vertically conformational variations, and variations of substituent position are arranged horizontally. The thickness of the surrounding lines represents approximately equal effectiveness, and the most effective compounds are in the middle. Compounds within the hatched lines are only slightly effective. The effectiveness is an expression of the stimulus quantity necessary to produce a distinct impulse activity (30 imp/sec within 200 ms of maximum response) averaged from 100 single cell recordings (Table 1).

Table 1.

Response classes of compounds	Average stimulus quantity to elicit 30 imp/sec
most effective	10^4
very "	10^5
medium "	10^6
slightly "	$10^7 - 10^{10}$
not "	$> 10^{10}$

In our discussion of the molecular properties in relation to the effect, we will start with the central and most effective group : 2-oxohexanoic acid, *trans*-3-hexenoic acid, and *trans*-2-hexenoic acid. Here we are dealing with an oxo group, a C=C double bond, the hydroxyl group and the carbonyl group. The presence of the double bond or the oxo group is not vital, if we can judge from the fairly effective hexanoic acid.

The same is true if one discards the hydroxyl group, as is shown by the equally effective unsaturated aldehydes. Absence of the carbonyl group leads to the alcohols which – when unsaturated – are markedly less effective. As already found by Boeckh (1967), it appears again that the saturation and the functional group can "support" each other. Effectiveness is not much influenced

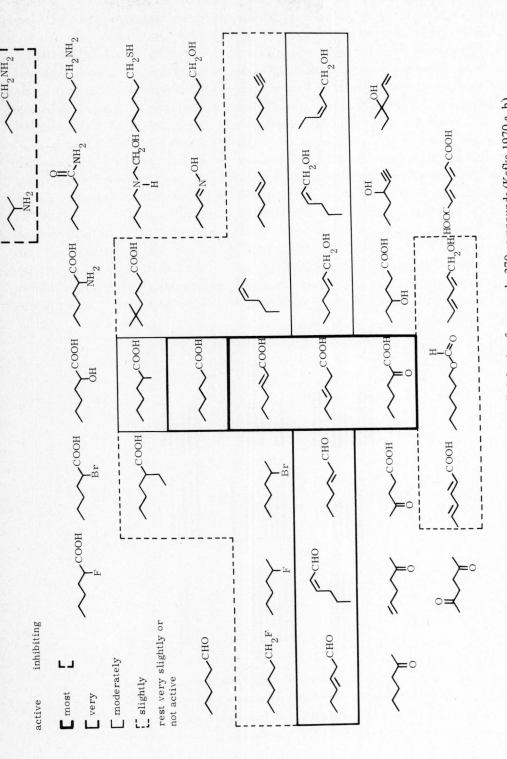

Fig. 3. Specificity of a single olfactory receptor cell. Selection from nearly 370 compounds (Kafka 1970 a, b).

by position and type of double-bond *(cis* or *trans) : trans*-2, *cis*-2 and *trans*-3-hexenal, as well as *trans*-2-, *cis*-2-, and *cis*-3-hexen-1-ol, respectively, are almost equally effective. However, in the only slightly effective unsaturated hydrocarbons the *cis*-conformation is more effective than the *trans*-conformation. *Cis*-2-hexene and *trans*-2-hexene are more effective than 1-hexyne.

It is of interest that the double bond in these hydrocarbons can be hydrohalogenated. Here, 2-bromohexane is more effective than 2-chlorohexane, and 2-bromohexane and 2-chlorohexane are somewhat more effective than the hexenes. In acids, however, hydrohalogenation leads to ineffectiveness. Addition of a second hydroxyl group or an amino group leads to a cancellation of the effectiveness, although short-chain amines may inhibit a spontaneously firing or an excited receptor cell. While one C=C double bond was essential for activity in alcohols and aldehydes, the introduction of a second C=C double bond diminishes the response. Compounds with more than one C=C double bond were always less effective. A similar reduction of the effect occurs when a second carboxyl group is introduced. In general, chirality was indispensable for a molecule to cause a cell reaction. None of these cells ever responded to a cyclic compound.

Comparing the shape of the compounds tested, one finds that this is not the essential principle for a molecule to be active. This is obvious when one compares the ineffective halogen or hydroxyl substitutions with the effective ones like the methyl or oxo substitutions. On the one hand, similarly shaped molecules (e.g. 2-oxohexanoic acid, 2-hydroxyhexanoic acid) may have different, and on the other hand, those having different shapes (e.g. 2-methyl hexanoic acid, *trans*-2-hexenal) may have identical effects.

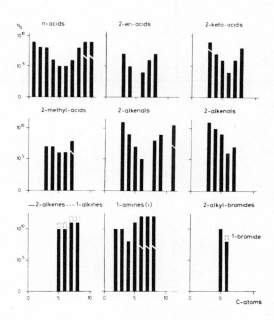

Fig. 4. Effectiveness of homologous compounds (Kafka 1970 a).
N_O = molecules per pit opening necessary to cause a reaction quantity of 30 imp/s or an inhibition (i) of an existing impulse activity by 20% (see text).

Variations of chain length are shown in fig. 4. In each class of homologs, molecules with a chain length of 6 carbon atoms proved to be most effective. On average, one methylene group more or less lowers the effectiveness by a factor of about 10.

LOCALISATION OF THE ACCEPTOR SYSTEM

Before looking for physicochemical correlations in active compounds one should know how the specificity is determined by properties of the sense organ. We are therefore looking for where the molecular interaction occurs and whether the Sensillen liquor or the electron-dense material interferes with the specificity.

A clear-cut answer came from the observation that two or more receptor cells can be immersed in the liquor while to a large extent responding independently. This would not be so if the liquor or the canal system acted as a specific filter. Consequently the dendritic membrane is the locus responsible for the specificity (for a precise discussion see Kafka 1970 a).

BINDING ELEMENTS BETWEEN ODOR MOLECULE AND ACCEPTOR SYSTEM

Because the system conducting the stimulus is not a specific filter, specificity can be described by the binding properties of the odor particles to definite acceptor structures in the receptor-cell membrane. If one considers the effect of chain length and position of C=C double bonds together with the effect of the terminal functional groups, it is obvious that this binding mechanism is of a complex nature. The first important deduction of the experiments is that covalent bindings are not involved in the transduction mechanism. This follows from a comparison of the cell response to related compounds : *trans*-2-hexenal (very effective), hexanal (ineffective) and *trans*-2-hexenol (moderately effective). If covalent forces were decisive, a classification of the responses should be possible based on aldehydes and alcohols respectively. Inspection of fig. 3 will reveal several more examples strengthening this argument.

How can we explain the complexity of the specificity? It is possible to describe the odor molecule acceptor interaction by two types of non-covalent binding elements. By "elements" we understand two parts of the odor molecule. One of them is largely defined by properties of the hydrocarbon structure, the other by properties of the functional groups. The first consists of a binding element which is not spatially directed because of the relatively free position and kind of C=C double bonds (*cis* or *trans*). The second one consists mainly of spatially directed binding elements. This follows from the fact that effectiveness was obtained only if the functional groups were located at the end of the molecule. But why are compounds with amino groups ineffective or even inhibitory? Submolecular binding dipoles in this case are oriented in an opposite direction to those of the hydroxy and carbonyl groups and this could be an explanation for the experimental observations.

The binding mechanisms between the hydrocarbon structure of the odor molecule and the acceptor system may be related to the ability to polarize the electron clouds of the intramolecular carbon bonds of the odor molecule. In this sense, interaction between hydrocarbon structure and acceptor system is based on **dispersion bonds** (binding type I). The binding of the terminal functional groups of an odor molecule to some polar (polar, since otherwise no positional effect of functional groups would be observable) or charged parts in the acceptor system consists mainly of **dipole-interaction** (binding type II). Hydrogen bonds could be involved in this binding process with the carbonyl group acting either as a **proton "acceptor"**, or the hydroxy group as a **proton "donor"**.

Binding elements (**polarisation bonds**) due to polarisation of the functional groups of the odor molecule (induced by the polar or charged parts in the acceptor system) also belong to this binding type II, too, as can be seen from the activity of the various halogenated hydrocarbons. Here, the effectiveness (not shown in fig. 3) is attenuated in the following way : 2-Br $>$ 2-Cl $>$ 2-F $>$ 2*cis* C=C $>$ 2*trans*-C=C. This sequence corresponds to the polarizability of these functional groups.

SUBSTRATE AND ACTIVE STRUCTURES OF ODOROUS MOLECULES

In analogy to the systems of enzyme chemistry I should like to suggest that the hydrocarbon part of the molecule indirectly reflects a substrate specificity and the part with functional groups reflects a reaction specificity of the acceptor system. I am inclined to speak of a "substrate" and of an "active" structure of the odor molecule. The substrate structure is responsible for loose contact to the acceptor (perhaps by solution) and the active one to a special site on it ("acceptor site").

TRANSDUCTION MECHANISM

Dipole interactions mostly occur by mutual alignment of the interacting partners. We can interpret the transducer mechanism as a sort of conformational change within the acceptor system caused by dipole interactions between odor molecule and acceptor system. This conformational change would then activate the reactive positions within the cell membrane (e.g. by allosteric effects) and cause a variation of the electrical permeability, finally leading to reaction of the cell. It is possible that similar mechanisms are responsible for the inhibition of the cells.

To the question of whether the inhibition of spontaneous or responsive cell activity is the result of a competition of two kinds of molecules for the same acceptor site we may reply negatively, at least for the spontaneous impulses, because in this case activating molecules are not present. However, the observation that excited cells could be inhibited only in certain cases, makes it probable that inhibition sites are not identical with excitation sites and leads to the idea of two sites on one receptor cell.

Gradation of effectiveness must not be understood to depend **only** on binding strength, because the continuous variations of response up to the factor 10^5 (as shown in fig. 3) are neither consistent with possible variations of the binding strength nor with corresponding binding effects. Therefore it seems that effectiveness is caused both by the probability that a molecule is able to reach the active sites and is also able to undergo specific adsorption.

The acceptor structures in the membrane must remain unaltered because (i), cell activity was reproducible even when using higher stimulus quantities and (ii), covalent bindings are not involved in the transducer mechanism for these odor molecules (cf. e.g. Kasang in this symposium). The logical consequence is that they are now transported to regions where they can no longer interact with the acceptors. They could, for instance, be transported into the cytoplasm of the dendrite.

The low olfactory threshold found in other organisms, such as the silk moth (Schneider, Kasang and Kaissling 1969, Kaissling and Priesner 1970), the eel (Teichmann 1959), the dog (Neuhaus 1956) and man (Stuiver 1958), and effects brought about by changing chain length, and type and position of the functional groups, makes it very probable that combined adsorption and binding of substrate and active structures to an acceptor system (as mentioned above) is the common principle for excitation of olfactory receptor cells.

We saw in the receptor for grass odor of locust that the odorants have only one active structure, which is essential for the final interaction. A much larger degree of specificity would be possible if more than one active structure were involved in the interaction mechanism. There are, perhaps, two active structures in bombykol (hexadeca-*trans*-10,*cis*-12-dien-1-ol), the sexual phero-mone of the female silk moth. Here, stereoisomers of the compound elicit much weaker responses (Schneider, Block, Boeckh and Priesner 1967; Priesner 1969, and Schneider, this symposium). One active structure could be due to the double bonds and the other to the alcohol group.

REFERENCES

Boeckh, J. 1962. Elektrophysiologische Untersuchungen an einzelnen Geruchsrezeptoren auf den Antennen des Totengräbers. *Z. vergl. Physiol. 46*, 212-248.

Boeckh, J. 1967. Reaktionsschwelle, Arbeitsbereich und Spezifität eines Geruchsrezeptors auf der Heuschreckenantenne. *Z. vergl. Physiol. 55*, 378-406.

Boeckh, J., Kaissling, K.E. and Schneider, D. 1965. Insect Olfactory Receptors. Cold Spring Harbor, Symp. on Quant. Biol. 30, 263-280.

Ernst, K.D. 1969. Die Feinstruktur von Riechsensillen auf der Antenne des Aaskäfers *Necrophorus* (Coleoptera). *Z. Zellforsch. 94*, 72-102.

Kafka, W.A. 1970 a. Analyse der molekularen Wechselwirkung bei der Erregung einzelner Riech-zellen. *Z. vergl. Physiol.* in press.

Kafka, W.A. 1970 b. Grundlagen der Spezifität einzelner Riechzellen. *Verh. d. Dtsch. Zool. Ges.* in press.

Kasang, G. 1970. This symposium.

Kaissling, K.E. and Renner, M. 1968. Antennale Rezeptoren für queen Substance und Sterzelduft bei der Honigbiene. *Z. vergl. Physiol. 59*, 357-361.

Kaissling, K.E. and Priesner, E. 1970. Die Riechschwelle des Seidenspinners. *Naturwiss. 57*, 23-28.

Lacher, V. 1964. Elektrophysiologische Untersuchungen an einzelnen Rezeptoren für Geruch, Kohlendioxyd, Luftfeuchtigkeit und Temperatur auf den Antennen der Arbeitsbiene und der Drohne (*Apis mellifica* L.). *Z. vergl. Physiol. 48*, 587-623.

Lacher, V. 1969. Ein neuer Sensillentyp auf den Antennen weiblicher Moskitos (*Aedes aegypti* L.). *Experientia 25*, 768.

Neuhaus, W. 1953. Ueber die Riechschärfe des Hundes für Fettsäuren. *Z. vergl. Physiol. 35*, 527-552.

Priesner, E. 1969. A new approach to insect pheromone specificity. Olfaction and Taste, C. Pfaffmann Rockefeller University press, 235-240.

Schneider, D. 1957. Elektrophysiologische Untersuchungen von Chemo- und Mechanorezeptoren der Antenne des Seidenspinners *Bombyx mori* L. *Z. vergl. Physiol. 40*, 8-41.

Schneider, D. 1968. Basic problems of olfactory research, in "Theories of odors and odor measure-ment", ed. Tanyolac, N. Technivision, England, 201-211.

Schneider, D. 1969. Insect olfacting : Deciphering system for chemical messages. *Science 163*, 1031-1037.

Schneider, D. 1970. This symposium.

Schneider, D., Block, B.C., Boeckh, J. and Priesner, E. 1967. Die Reaktion der männlichen Seiden-spinner auf Bombykol und seine Isomeren. Elektroantennogramm und Verhalten. *Z. vergl. Physiol. 54,* 192-209.

Schneider, D., Kasang, G. and Kaissling, K.E. 1969. Bestimmung der Riechschwelle von Bombyx mori mit Tritium markiertem Bombykol. *Naturwiss. 55,* 395;

Steinbrecht, R.A. 1969. Comparative morphology of olfactory receptors. Olfaction and Taste. ed. C. Pfaffman Rockefeller University Press, 3-21.

Stuiver, M. 1958. Biophysics of the sense of smell. Diss. Math. Nat. Fak. Groningen.

Teichmann, H. 1959. Ueber die Leistung des Geruchssinnes beim Aal. *(Anguilla anguilla* L.). *Z. vergl. Physiol. 42,* 206-254.

DISCUSSION TO TALKS OF SCHNEIDER AND KAFKA

Randebrock : The path of a molecule on its way to the receptor sites can be thought of as a series of steps :passage of a membrane, for example, followed by solution in a liquid, then actual fitting to the receptors. Perhaps the situation concerning specificity therefore arises from changes in one or other of these steps.

Kafka : I considered this point and found specificity is, on the whole, not brought about by specific filter systems but by specific interaction within the region of the cell membrane. Experimental evidence in support of this is that I was able to cause inhibition of a cell by both hexanal and hexenal, while another cell in the **same** liquor was inhibited only by hexenal.

Kovats : You observe a spike at a certain concentration, but I notice that as you lower the concentration the signal is delayed. Can this mean that the antennae concentrates the substance until sufficient is available for the signal?

Schneider : A "concentration" of stimulating molecules can be excluded in our case because at threshold, only one per cent of the cells receive **one** molecule each per 0.86 sec (the average behavioral reaction time). There is neither time, nor a morphological basis for a concentration of the randomly hitting molecules. The latency (i.e. the average time difference between the molecule impact and the impulse) is approximately 0.5 sec for stimuli up to 10^{-3} μg of bombykol. The latency shortening with higher stimulus intensities indicates integrative processes in the receptor cell. At this stage the cell does not "count" the molecules any more.

Kalmus : Did you have any spikes without stimulation of the receptors?

Schneider : Yes, this is what I called "noise", the cell fires on the average a spike to the brain every ten seconds.

Hughes : Kafka has shown that some excitatory sites are not the same as the inhibitory sites. In Schneider's diagrams of the impulses, I was surprised to see a discontinuity, and I wondered if you could explain this by simultaneous activation of excitatory and inhibitory points, the point where the impulses stop being when the inhibition starts breaking through.

Schneider : The discontinuity of the impulse distribution at low stimulus intensities is probably an expression of the randomly hitting molecules. With stronger stimuli, this discontinuity disappears. We have not the impression that the reaction of the bombykol receptor cell needs or favors an explanation where excitatory and inhibitory points are at an interplay with one another. The inhibition which Dr. Kafka mentioned was observed with odorants different from the exciting ones.

Mann : As a perfumer I have encountered an interesting facet of the sensitivity problem. We were trying to detect the threshold at which an odor change of a complex perfume mixture was perceptible after addition of a macrocyclic musk (e.g. pentadecanolide, about the same volatility as bombykol). At a dilution of 10^{-17} this was still perceptible. After computing the vapor pressure of the musk we concluded that the amount entering our nose was of the order of several hundred molecules per second, meaning that possibly several dozen per second were effective at the olfactory nerve endings. I made a guess some years ago that olfactory mechanism might be comparable to that of the eye, which uses isolated photons, and the insect results appear to support this.

TASTE SENSATIONS AND EVOKED BRAIN POTENTIALS AFTER ELECTRIC STIMULATION OF TONGUE IN MAN *

K.-H. Plattig
I. Physiological Institute, Erlangen.

The classical tool of physiology, namely the performance of experiments on living animals or with dissected surviving tissues or organs appears blunt when applied to problems of sensory physiology. Thus while physiology of respiration, of circulation, of kidney or of muscle is an "objective" science, sensory physiology has the disadvantage of depending on indirect, more time-consuming or more expensive behavioral studies in animals for objectivation of sensations or on subjective, psychophysically gained explanations of humans.

This might have been the reason,in part that studies on taste, for example, have taken quite a long time to reveal anatomically the central terminals of gustatory pathways in man. The method of recording evoked potentials from the brain of an awake man through his intact skull after sensory stimulation appeared to be of help. This method had been developed by Dawson (1947), who used repetitive stimulation of somesthetic afferents and superimposed fifty successive EEG sweeps optically on an oscilloscope, each starting with the precise moment of stimulus. This method was improved by using modern computers for summating or – in more common language – averaging the EEG, as elaborated by Keidel et al. (1965) for the acoustically evoked potential and for other modalities (Fig. 1).

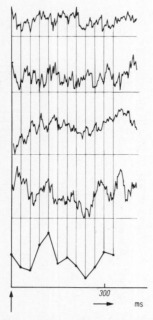

Fig. 1. Method of summating or "averaging" post-stimulus sweeps of EEG to
disclose the hidden evoked potential shown in the lowest track.
Drawing by M. Spreng from W.D. Keidel (1965).

*Dedicated to Prof. Dr. H. Kiese on the occasion of his 60th birthday.

The technique of averaging the EEG has one precondition however, which can hardly be fulfilled by rinsing the tongue with chemical solutions for physiological stimulation of taste : the moment of stimulation has to be fixed within a range of less than 1 millisecond to obtain a clear cortical answer. This holds for the use of averaging computers as well as for optical superposition, and certainly for manual evaluation too, the latter taking at least 1,000 times longer than any other method.

The only way round appeared to be by using inadequate electric stimulation of the tongue, which certainly has the disadvantage of possibly mixing various different sensations of somesthetics (thermal, touch, pressure, or even pain) among taste. We will refer to this difficulty later, but initially we felt encouraged to use "electric taste" which had been known since the days of Volta or even Sulzer. In the 200 years from Sulzer's first experiments, electric taste was examined very thoroughly by a number of scientists, of which I would like to quote Ohrwall as the most important of the older ones. Of the present generation the Nobel-Laureate of 1961, von Békésy (1964), should be mentioned together with Bujas (1937), and Dzendolet (1957, 1962). Dzendolet (1957) had been the first to use electrical rectangular pulses of different but precise duration and frequency, and this method was used also by v. Békésy (1964), who succeeded in getting all the four taste qualities, sweet, sour, bitter and salty by modifying the frequency and independently the duration of the electrical pulses, which were applied over a rather large silver electrode of 70 mm^2 to the upper antero-lateral tongue edge. We used this method first for recording averaged evoked brain potentials from the parietal lobe projection (Gyrus postcentralis) versus contralateral frontal area (Jauhiainen 1966; Plattig, 1969). These potentials were present only in subjects who in the same experiment also reported subjective taste sensations; they were not present when the taste threshold was raised by adaptation, e.g. by strong spices like curry or vinegar taken less than one hour before the experiment was performed. Fig. 2 shows the experimental device as a block diagram. Three

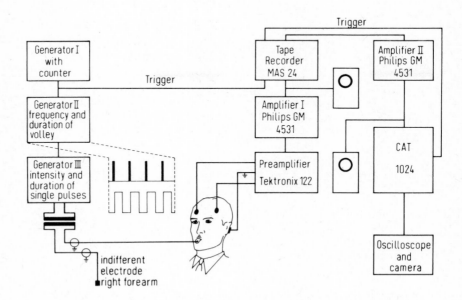

Fig. 2. Experimental device as block scheme. Details in text. The coupling time
constants of the preamplifier were above 0.2 seconds, while the amplifier
(Philips GM 4531) is DC-coupled.

pulse generators are in series; the first one has a counter and triggers the second, and the second triggers the third, enabling the production of volleys of pulses, the frequency and duration of which could be varied independently. The whole volley was automatically terminated at a certain variable time after the first pulse was started, the latter also being used as trigger pulse for the averaging computer into which the amplified and tape-stored EEG was fed. The positive pole of the third generator was linked to the tongue electrodes shown in Fig. 3, across a stimulus isolation unit of our own construction while the negative pole was linked with the subject's forearm. To improve these stimu-

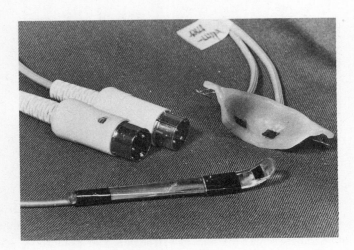

Fig. 3. Electrodes used for stimulation of tongue. Palatal electrode held with plugs (above) consisting of "Paladur R" and handle electrode (below) consisting of plexiglass. The electrodes themselves are 0.2 mm silver sheets, with sizes of 6, 10, 20 or 70 mm^2. Silver-silver chloride electrodes used in the beginning did not give noticeable differences, so that in general stimulation was done by plain silver. From K.-H. Plattig (1969).

lating facilities, together with R. Rix (1969) and with technical assistance of G. Witzel, we recently constructed an "electrogustometer" in which the functions of all the three generators, including the stimulus-isolating unit, are housed together in one box as shown in Fig. 4, 5 and 6.

After obtaining the first records of brain potentials we had to obtain more psychophysical values on electric taste by our method, and we proceeded first by studying intensity functions together with J. Helmbrecht (1968) and secondly by studying quality relations.

Examinations of sensory intensities as functions of stimulus intensities were first undertaken by E.H. Weber and C.T. Fechner, whose well-known law, established in 1834, describes the dependence of difference-thresholds on logarithms of stimulus intensities in the medium intensity ranges. It was attacked by S.S. Stevens (1957, 1960) at Harvard University over the last 30 years, who postulated, instead of Weber-Fechner's logarithmic function, a power function of the somewhat simplified form, $R = S^n$, in which R represents intensity of sensation, S intensity of stimulus and n an exponent characteristic for each sensory modality under defined stimulus conditions. While for a number of single sensory elements logarithmic functions have been recently reestablished, Stevens and many of his co-workers succeeded in proving these power-functions for quite a number of modalities and stimulus conditions, and we used his "ratio-production-method" for electric taste too.

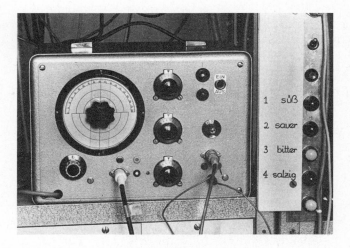

Fig. 4. "Electro-Gustometer" as stimulation unit with facilities to vary frequency between 2 Hz and 20,000 Hz, duration of single pulse between 0.01
and 10 ms, duration of volley between 2.5 ms and 22 s and amplitude
of pulse between 10 mV and 20 V. The pulses are produced as symmetrical deflections from the zero line, so that no considerable influence
on EEG is caused. Lights for signaling the quality sensed are posted on
the right. From Plattig and Rix (in preparation).

Fig. 5. Interior of the "Electro-gustometer". From Plattig and Rix (in preparation).

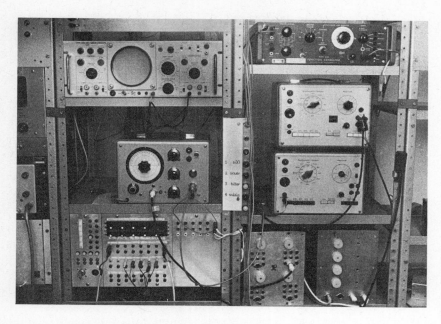

Fig. 6. The whole stimulus unit. On the right the three generators as used previously;
left side : oscilloscope (top) to monitor the stimuli, the "Electro-Gustometer"
(middle), and switching devices for access to a Linc-8 computer.

Fig. 7 shows one of the subjects in an electrically and acoustically shielded room with the electrode
in position on the tongue. An attenuator was interfaced between the gustometer and the electrodes,
enabling the subject to decrease or increase stimulus intensity in decibels. With fixed parameters of
pulse duration and frequency, a certain intensity was offered to the subject for 3 to 5 seconds, when
he had to tell the supervisor to change the attenuator position in such a way that sensation intensity
doubled. To avoid disturbances by adaptation all experiments were done between 8 and 11 a.m. and
all subjects were requested not to smoke or to take sharp spices after 8 p.m. the day before. We
doubted that human discrimination could be fine enough to give reproducible results, and were surprised and pleased to find that with logarithmic coordinates, straight lines characteristic of power
functions as described above were obtained.. Using the Bravais-Pearson correlation coefficient the
power function gave a better fit than a logarithmic function, and the overall exponent for various
pulse durations and frequencies was 1.075 ± 0.17, as shown in Fig. 8, together with the functions
of other single combinations of pulse durations and frequencies. Our values for the exponents of
electric taste fitted well with that of 1.054 found for chemical stimulation by Beebe-Center and
Waddell (1948) and by D.R. Lewis (1948) of the Stevens'group (Fig. 9), and we will use them later
for determination of specifity of the evoked brain potentials. I should mention here that if higher,
but individually varying electric intensities are applied to the tongue, taste sensations are displaced
by the sensation of electric shock or even of severe pain. This sensation of electric shock had a
higher exponent of 1.95 ± 0.11.

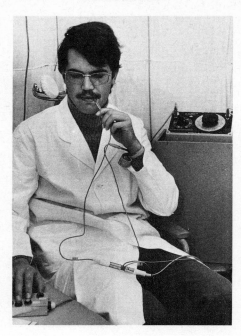

Fig. 7. Subject in the shielded chamber during a test with handle-
electrode while pressing push-buttons to signal qualities
sensed. The attenuator in the rear facilitates varying of
stimulus intensity.

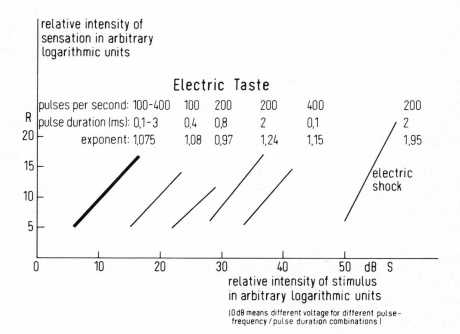

Fig. 8. Exponents of the power functions of electric taste for different combinations of frequency
and duration of stimulating pulses in comparison with the exponent of electric shock.
Details in text. From J. Helmbrecht.

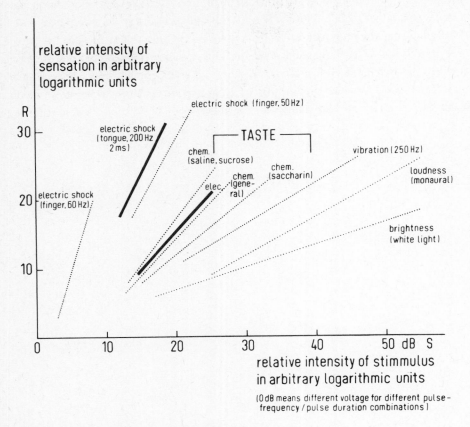

Fig. 9. Stevens's exponents for various modalities compared with our exponents shown in Fig. 8. From J. Helmbrecht.

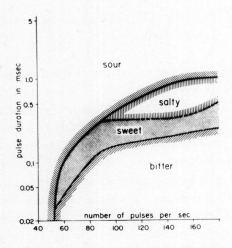

Fig. 10. "Quality areas" after electric stimulation of tongue dependent on the number of pulses per second (abscissa) and pulse duration (ordinate). Details in text. From G.v. Békésy.

The next step should be a clearer understanding of the quality relations. Von Békésy (1964) has published a diagram (Fig. 10) in which he plotted pulse duration versus frequency. When stimulating a rather small number of (Asian) subjects successively with a large number of stimuli with frequency-duration parameters marked by a dot in this figure, he added the qualities (sweet, sour, bitter or salty) sensed at each dot, thus revealing fields of qualities. We have just completed the systematic evaluation of the same kind of experiment by the χ^2-test on 16 male and 5 female subjects which showed that for (German) students of psychology or medicine the distribution was different from von Békésy's results (Fig. 11). All subjects were stimulated

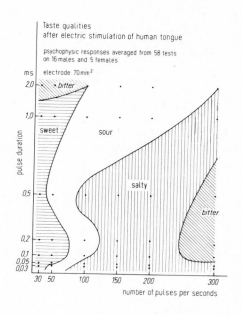

Fig. 11. "Quality areas" of 21 German students. Details
in text and Fig. 10.

in the same way as described above, in at least 7 sessions of about one hour's duration each. For each session one fixed pulse duration (0.03, 0.05, 0.1, 0.2, 0.5, 1 and 2 ms) was chosen statistically, and 6 frequencies (30, 50, 100, 150, 200 and 300 pulses per second) were offered in highly random order with an intensity previously fixed at the individual optimum. The pulses were offered for 15 seconds, and during the next 15 seconds the subject had to report by visual signal from the shielded chamber what kind of sensation occurred. Although each subject was trained by chemical solutions before starting the experiment, gustatory discrimination for electrical taste improved clearly during the first three or four sessions. From about 150 sessions we used 58 of high significance ($p < 0.01$) for the fields of qualities shown in Fig. 11. We also found that the differences between individuals were considerable, so that each point in Fig. 11 gives only a certain higher probability that the quality indicated will be sensed. We did not examine individual quality and threshold variations in respect to the personal psychological taste characteristics as described by R. Fischer et al. (1965), but we shall try to do so in future.

Now with a somewhat better understanding of the subjective events after electrical stimulation of tongue, we should examine the evoked brain potentials, work that we had started some time ago as I have already mentioned. I would remind you of Fig. 2, showing the device for electric stimulation of the tongue and for recording and processing the EEG. At least 40 well-processible post-stimulus EEG sweeps had to be available for averaging. This means that at least 50 to 60 stimuli had to be presented, 10-20 of which had to be omitted on account of their being technically inadequate, e.g. for moving artefacts. To minimize adaptation, the stimulus-volleys were given every 20 seconds, each being of 30 ms duration. For most of the experiments with brain potentials,we have so far used a pulse duration of 2 ms and a frequency of 200 pulses per second, so that each volley consisted of 5 pulses. The experiments were also done between 8 and 11 a.m. to avoid influences of adaptation and of individual fatigue of the subjects, and all subjects were requested not to smoke nor to take strong spices from at least the preceeding evening. 10 subjects were examined in 453 sessions with 12 stimulus intensities.

Brainpotentials after electric stimulation of tongue

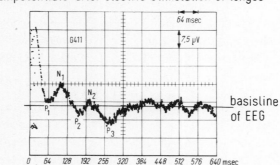

Fig. 12. Typical brain potential after electric stimulation of
tongue. Details in text. From K.-H. Plattig (1969).

Fig. 12 shows one characteristic brain potential gained in this way with the recording EEG-electrodes (3.5×5 cm^2 silver with saline sponges) over parietal lobe versus frontal area. One can see the initial stimulus artefact which would have displaced the EEG beyond the scope if it had not received symmetrical deflections by the stimulus isolation unit. Recording from vertex versus chin or mastoid enlarged the deflections. Negative (N_1, N_2) and positive (P_1, P_2, P_3) amplitudes can be seen, the peaks of which were measured from a base-line of EEG being extrapolated from the "quiet" EEG-sweep before and after the response. The latencies of the amplitudes decreased with increasing stimulus intensities, the ranges in ms being shown in Fig. 13. A difference in amplitudes or latencies

P_1	N_1	P_2	N_2	P_3
45–80	90–155	175–195	210–245	285–420 msec

Fig. 13. Table of latency ranges of the various deflection of brain potentials. P for positive and N
for negative deflection under the parietal electrode. From K.-H. Plattig (1969).

Brain potentials after electric stimulation of tongue

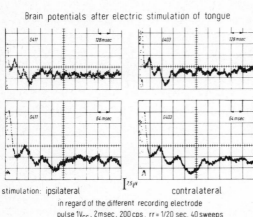

stimulation: ipsilateral   contralateral
in regard of the different recording electrode
pulse $1V_{SS}$, 2msec, 200 cps, rr = 1/20 sec, 40 sweeps

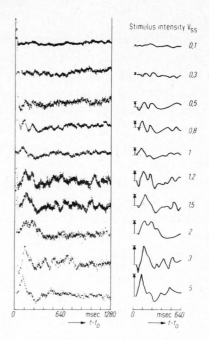

Fig. 14. Brain potentials recorded from the ipsi- and contra-lateral parietal area in regard to position of tongue electrode. From K.-H. Plattig (1969).

Fig. 15. Increase of amplitudes of brain potentials with stimulus increase. From K.-H. Plattig (1969).

could not be seen for stimulating the ipsi- or contralateral edge of tongue in reference to the different (parietal) EEG-electrode (Fig. 14). The amplitudes N_1 and P_3 increased characteristically with increasing stimulus intensity (Fig. 15), and followed a power function of stimulus intensity with very high correlation, as can be seen in Fig. 16. This dependence holds also for the peak to

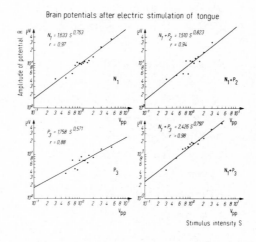

Fig. 16. Intensity functions of the amplitudes N_1, P_3, N_1+P_2 and N_1+P_3 of brain potentials after electric stimulation of tongue. The power-function as listed in each area together with the correlation-coefficient, r, were worked out by a Linc-8 computer. From K.-H. Plattig (1969).

peak amplitudes of N_1+P_2 and N_1+P_3. The exponents of the function of N_1, N_1+P_2 and N_1+P_3 range between 0.763 and 0.823, but for P_3 it is much smaller, being 0.571 (Fig. 16). The remaining amplitudes did not give any correlation (Fig. 17).

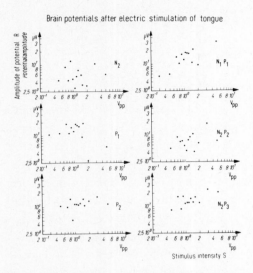

Fig. 17. The amplitudes N_2, P_1, P_2 and the combinations N_1+P_1, N_2+P_2 and N_2+P_3 did not show a dependence on stimulus intensity as shown in Fig. 16. From K.-H. Plattig (1969).

There are three statements which lead us to believe that these amplitudes with exponents around 0.8 are correlates of taste sensation in spite of the difficulties mentioned above by inadequate stimulation :

1. The amplitudes were present only when subjective taste sensations were reported.

2. They did not occur when a spot of tongue, lips or cheeks was stimulated, where taste papillae were absent, or when the tongue had been treated by tetracaine or by curry (Fig. 18).

3. The exponent of the intensity function multiplied by about 1.6 gave the value of the exponent obtained psychophysically for the same modality. This is the same factor of 1.6 to 1.7, which in earlier examinations Keidel and his group had found for this relation between "objective" exponents of brain-potential functions and exponents of psychophysical functions, e.g. in hearing, vision, vibration and pain.

Two points remain to be considered. The first is the meaning of the factor 1.6 to 1.7, mentioned in statement 3. The other is the question of future work.

Concerning the factor of nearly 1.7, our initial impression proved incorrect. Stevens, Mack and Stevens (1960) had the brilliant idea of scaling subjective intensities by "cross-modality matches". Matching the intensities of subjective sensations of different sensory modalities means expressing, for example, the magnitude of acoustic sensation by the brightness of a lamp which can be modified by varying a resistor in the powerline leading to the lamp. In the same way acoustic and gustatory sensations can be matched; taste-sensation magnitude could be expressed by loudness and vice versa.

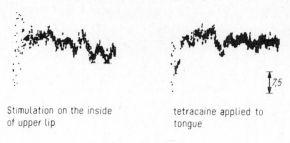

Stimulation on the inside   tetracaine applied to
of upper lip   tongue

1Vpp, 2ms, 40 sweeps

Fig. 18. Stimulation of mouth areas without taste papillae or
with anesthetized taste cells (does not reveal potentials).
From K.-H. Plattig (1969).

Stevens's group preferred handgrip or muscle-sense (kinesthesis), and they used their famous hand-grip-method by pulling a dynamometer to match a wide variety of sensory continua with kinesthesis by force of handgrip. The sensation R_1 produced by the force of handgrip is expressed by the power function $R_1 = S_1^m$, in which S_1 is the stimulus intensity for the subcutaneous mechano-receptors; this stimulus intensity is expressed by the "force of handgrip" in pounds or grams, and is measured directly on the scale of the dynamometer. The exponents m of this function was found to be 1.7, and that is the reason that we started thinking about it. If there is another modality with the respective power function $R_2 = S_2^n$, and if the subjective sensation of this modality is to be expressed by the magnitude of the force of handgrip, the consequence is that R_1 should be made equal to R_2. The following equation is therefore valid : $S_1^m = S_2^n$, or $S_1 = S_2 \frac{n}{m}$. This means that an exponent given by a direct scaling method has to be divided by $m = 1.7$ to have the expo-nent for the same modality as revealed by the handgrip method.

Since all the averaged brain potentials of different modalities show exponents which are about equal to those obtained psychophysically by the handgrip method, we felt that some inter-action between muscle-sense and the modality examined might be the reason for this characteristic reduction of the exponent, but this was not the case, and we have to admit that we do not yet know the reason.

Concerning the second question raised above as to future work, we are trying to compare responses of single elements at different levels of the gustatory pathways when the tongue papillae are excited chemically or electrically. We now think that precise triggering of an averaging compu-ter after stimulation of single papillae either electrically or by solutions from micropipettes might become possible not too far in the future, and we hope then to obtain more and better knowledge of information processing in the gustatory pathways, as well as of electric taste.

REFERENCES

Beebe-Center, J.G. and Waddell, D. 1948. A General Psychological Scale of Taste. *J. Psychol. 26,* 517-524.

Békésy, G. von. 1964. Sweetness produced electrically on the tongue and its relation to taste theories. *J. appl. Physiol. 19,* 1105-1113.

Bujas, Z. et Chweitzer, A. 1937. "Goût électrique" par courants alternatifs chez l'homme. *C.R.Soc. Biol. 126,* 1106-1109.

Dawson, G.D. 1947. Cerebral responses to electrical stimulation of peripheral nerve in man. *J.Neurol. Neurosurg. Psychiatr. 10,* 134-140.

Dzendolet, E. 1957. Intensity-duration relation for taste using electrical stimulation. Thesis, Brown Univ., Providence, R.I.

Dzendolet, E. 1962. Electrical stimulation of single human taste papillae. *Perceptual Motor Skills 14,* 303-317.

Fischer, R., Griffin, F., Archer, R.C., Zinsmeister, S.C. and Jastrau, P.S. 1965. Weber ratio in gustatory chemoreception, an indicator of systemic (drug) reactivity. *Nature (Lond.) 207,* 1049-1053.

Helmbrecht, J. 1968. Psychophysik des elektrischen Geschmacks : Qualitäts- und Intensitätsbeziehungen. *Arch. klin. exp. Ohr.-Nas.-u. Kehlk.-Heilk. 192,* 614-617.

Jauhiainen, T. und Plattig, K.-H. 1966. Reizsynchrone langsame Rindenpotentiale beim Menschen nach elektrischer Reizung der Zunge. *Pflügers Arch. ges. Physiol. 289.* R. 27 (Abstr.).

Keidel, W.D. 1965. Neuere Ergebnisse der Physiologie des Hörens. *Arch. Ohr.-Nas.-Kehlk.–Heilk. 185,* 548-575.

Lewis, D.R. 1948. Psychological Scales of Taste. *J. Psychol. 26,* 437-446.

Plattig, K.-H. 1969. Ueber den elektrischen Geschmack. Reizstärkeabhängige evozierte Hirnpotentiale nach elektrischer Reizung der Zunge des Menschen. *Z. Biol. 116,* 161-211.

Plattig, K.-H. und Rix, R. 1969. Rechteckimpulsgenerator zur Auslösung subjektiver Geschmacksempfindungen definierter Qualität. *Pflügers Arch. 312,* R 128 (Abstr.).

Stevens, J.C., Mack, J.D. and Stevens, S.S. 1960. Growth of sensation on seven continua as measured by force of handgrip. *J. exp. Psychol. 59,* 60-67.

Stevens, S.S. 1957. On the psychophysical law. *Psychol. Rev. 64,* 53-181.

DISCUSSION

Zottermann: We started about 15 years ago looking for cells in the tongue area of the cat responding to chemical stimulation, but the results were very poor. Later on we found that stimulation of the gustatory nerve leading to the anterior part of the tongue, produces an evoked response which is a composite signal consisting of a first quick wave followed by a later wave. The second wave only is characteristic of taste. Taking advantage of patients operated for otosclerosis, we were able to fix electrodes in the tympanum in the middle ear. Thus we measured evoked electric response to chemical stimulation.

In the 6-7 cases studied so far, we have found absolute congruity between perceptual responses and the summated electric responses.

Kalmus : I should like to point out in connection with Plattig's and Zottermann's work, that in gustation there are two types of people, the phenylthiocarbamide (PTC) tasters and the PTC non-tasters.

Zottermann : We have found that PTC non-tasters give us response and although we have not yet done the experiment, we are fully convinced that PTC tasters will respond. Specific sensitivity to one taste might well be genetically controlled.

Hughes : I should like to comment on three phenomena which might interfere with Plattig's frontal reference electrode :

1) The phenomenon of contingent negative variation which occurs over the frontal region.
2) The possibility of picking up small pulses in connection with eye movements.
3) The possibility of hypersynchronization of alpha activity.

Plattig : We paid much attention to the first point. The rhythm of intervals between the stimuli was not kept strictly at 20 seconds, but varied between 15 and 25 seconds, in order to exclude the possibility of getting a negative contingent variation.

The second point concerning the so-called Bickford effect was also considered. It occurs within 30 to 50 ms after stimulus onset, and is hidden by the stimulus artefact in our experiments. In future experiments, however, recordings will be made from the vertex versus either the chin or possibly the mastoid. Preliminary experiments indicate that the results are qualitatively exactly the same, but amplitudes of these vertex recordings are higher.

Finally, concerning the third point, we cannot decide at present whether alpha synchronization or over-synchronization is involved or not.

Fischer : I should like to stress the fact that information and meaning must be clearly separated. Meaning is interpretation of information and man is a self-referential system. Without reference the same information will have different meanings for different people because they do not know how to interpret it. On the other hand you may change the information, e.g. by putting on color-distorting glasses. After a fairly short time, people will have adapted to the modified information and see the world as it should be.

Plattig : In terms of cybernetics and information theory, this is referred to as "information" (indicating a quantum of bits) and "code" (indicating the quality or the meaning).

WHAT MAN'S NOSE TELLS MAN'S MIND

A COMPARISON BETWEEN SENSORY SCALES OF ODOR INTENSITY AND THE ELECTRO-OLFACTOGRAM IN MAN

O. Franzén, P. Osterhammel, K. Terkildsen and K. Zilstorff
Department of Neurology, Karolinska Sjukhuset,
Stockholm 60, Sweden.

"Nihil est in intellectu,
quin prius fuerit in sensy."
Giovanni Nevizano,
Sylva nuptialis, 1521

Two lines of research have existed side by side in the study of sensory systems. The psychophysicist deals with the over-all performance of an organism whereas the neurophysiologist explores the functioning of the nervous system by tapping the afferent neurons of the sensory system. Most experiments in neurophysiology have been carried out on animals. In spite of the great value of such experiments there are certain inherent limitations. It is by no means apparent that results from animal experiments can be directly applied to man. For the students of the olfactory system the frog has been the pet animal, but unfortunately we cannot ask the frog what he thinks about an odorant.

The quantification of subjective variables plays an increasingly important role in modern psychology and research carried out in recent years bear witness to the success of this approach (Ekman & Sjöberg 1965). The direct scaling methods of modern psychophysics originate from the work of Stevens (1957) who demonstrated that sensory magnitude (R) grows as a power function of stimulus intensity (S), $R = cS^n$ or, $\log R = \log c + n \log S$. Under the method of magnitude estimation the subject is required to make quantitative estimates of sensory magnitudes, or in other words he is asked to assign a number proportional to the perceived intensity of the physical stimulus. The first application of the method of magnitude estimation to the problem of perceived odor intensity was made by Jones (1958 a, b). For various chemical compounds he obtained power functions with exponents between about 0.4 to 0.6. This psychophysical power function is an empirical law and its form might be explained in neurophysiological terms.

At the present time, to our knowledge, only two studies have successfully achieved a bridge between sensory experience and human neuroelectrical events (Borg, Diamant, Ström & Zotterman 1967; Franzén & Offenloch 1969). Borg et al. (1967) recorded the summated neural response of the chorda tympani in the middle ear to different gustatory stimuli. The same stimuli were rated using the method of magnitude estimation. They found that both the subjective and neuroelectric data were described by power functions and that the two sets of responses were highly correlated. Franzén & Offenloch (1969) observed that computer-averaged primary potentials recorded from the somatic receiving area of the postcentral gyrus exhibited a close correspondence with sensory scales, i.e. the cortical responses and the magnitude estimates were power functions of the tactile stimulus with exponents of approximately the same value. In the investigation to be reported here we have extended this line of work following essentially the same experimental and theoretical approach.

The first link of the transmission chain from the periphery to the brain is the receptor organ. The sense of smell offers a rather unique opportunity since its sensory epithelium is accessible in man – though not easily – without surgical incision. As yet no direct evidence concerning the relation between the transducer process in the sense organ and perceived intensity has been brought forward. Information about this relationship is of utmost importance since there are reasons to believe that the input-output characteristics of the transducer determines the fundamental relation between physical intensity and sensation magnitude. The transducer elements of the sense of smell are bipolar cells embedded in the olfactory mucosa which takes up an area of about 2.5 cm^2 in the upper part of the nasal cavity on each side of the septum in man. The cilia penetrate through a layer of mucus covering the sensory cells. On good grounds it is assumed that these hairs are the locus of stimulus energy transduction. The axons of the sensory cells pass to the bulb through the cribiform plate. From the bulb fibers go to the prepyriform area of the brain. The cerebral connections are not well understood.

A study of the manner in which odorants are processed by the olfactory system can provide the basic information to define a measure of intensity. The electrical changes following olfactory stimulation can be studied with a macroelectrode placed on the mucosa. The response of the olfactory epithelium was discovered by Ottoson (1956). He introduced the concept of electro-olfactogram (EOG) in analogy with electroretinogram (ERG) since the two receptor potentials possess several common features. Ottoson's pioneering work on the frog constitutes the foundation for the interpretation and evaluation of the findings of the present study. All experimental evidence supports the view that the EOG is a pure receptor potential. The method of recording the EOG has been available for more than a decade but not until now it has been possible to relate these data to corresponding measurements in man. When stimulated with an odorous substance the receptor cell produces a local current. This current spreads to the nerve and triggers an impulse train which carries the information of the stimulus to the brain.

Sniffing is the most efficient way of smelling as it sets up vigorous eddying currents of air in the nasal cavity. In the present study sniffing was not desired during the recordings and therefore a modified form of Elsberg's blast injection technique (Elsberg & Levy 1935) was considered an acceptable compromise. For stimulation we used coffee-saturated air. The odorous air was accordingly injected into one nostril of the subjects under voluntary apnoea and its volume was varied between 2 to 25 cc. We recorded the EOG by placing a chlorated silver electrode on to the olfactory region and maintained the electrode in a position that gave the maximal baseline-to-peak response. There were great difficulties in introducing the electrode without causing mucuous discharge, sneezing and distraction. A typical receptor potential is displayed in Fig. 1 in response to 25 cc coffee odor. An upward deflection signifies a negativity of the active electrode. The response is characterized by a rapidly rising phase and an exponential fall toward the baseline. It is important to note that there is no off-response. Room air evoked no potential or a potential of negligible magnitude. Our potentials are identical with the EOG in the frog except for the time course, which is faster in man. The bar on the response tracing indicates onset of the stimulus. This does not permit a determination of the true latency as we do not know the time it takes for the particles to pass from the outlet of the nose to the sensory cells.

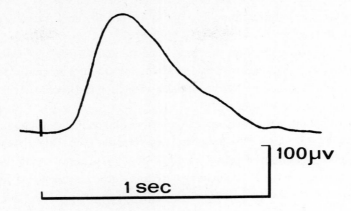

Fig. 1. Receptor potential response to coffee odor.

The EOG is a mass response that does not allow us to discern the unique properties of the individual cells. The recorded potential may, however, be assumed to reflect the activity of a representative sample of cells. This way of recording receptor activity is therefore an advantage rather than a limitation, bearing in mind the complexity of a mixture of odorous compounds such as is present in coffee odor.

As the peak amplitude of the EOG is plotted in log-log coordinates a straight line describes adequately the data with a slope ranging between 0.49 to 0.55. The EOG seems thus to be a powerful tool for a quantitative assessment of the magnitude of the stimulus. As a matter of fact this is the first organ in which the functional relationship between the generator potential in man and a physical continuum is demonstrated. The results of the magnitude estimations indicate that the individual data are satisfactorily described by a power function. The mean exponent was 0.52 and the standard deviation ± 0.14. The variation observed in the individual data, we believe, is a rather typical outcome of numerical scaling experiments of perceived intensity of odor. The biophysical and the average psychophysical exponents differ only in the second decimal place. It is of particular interest that Reese and Stevens (1960) have reported a group exponent of 0.55 for coffee odor when concentration was used as the independent variable.

Perceived odor intensity under short-term stimulation is thus found to be a power function of either volume, V, (concentration being kept constant) or concentration, I, (volume being kept constant). The mechanism underlying these effects can presumably be attributed to an increase in receptor response amplitude and to a greater number of sensory cells activitated. Different volumes and different concentrations of butanol, the product of volume and concentration being kept constant, elicited an EOG of equal amplitude in frog (Ottoson 1956). Olfaction might share an important property as to energy integration with another sensory system, namely vision. The reciprocity for vision, expressed in the Bunsen-Roscoe law, $I \times T = C$, where T stands for time, is also only valid for brief presentations of stimuli (Hartline 1928). A mathematical consequence of the reciprocity, $I \times V = C$, for olfaction with respect to the intensity function would be $R = k(I \cdot V)^p$ or $\log R = \log k + p (\log I + \log V)$. As can be seen, concentration and volume should be raised to the same power.

The observations of Reese & Stevens (1960) and those of the present study are consistent with this notion, although they should not be accepted without reservation since we are faced here with great problems as to control and appropriate application of the stimulus. They do, however, suggest future studies of great importance.

The close correspondence between receptor activity and sensory experience suggests that the information from the transducer is signalled to the brain without much distortion as far as intensity is concerned and that the form of the magnitude function is determined already by the response characteristics of the sensory cells.

The present findings are in harmony with previous observations on action of synapses (Eccles 1964; de Reuck & Knight 1966), peripheral (Mountcastle 1966; Granit 1966; Borg et al. 1967) and central neurophysiology (Franzén & Offenloch 1969). The power law transfer function may represent an important mode of action of sensory transducers. Invariance of the sensory information is preserved within the nervous system by linear transformations. These two general functional principles point to a very simple and equally attractive mechanism of intensity coding in sensory systems.

REFERENCES

Borg, G., Diamant, H., Ström, L. and Zotterman, Y. 1967. The relation between neural and perceptual intensity : a comparative study on the neural and psychophysical response to taste stimuli. *J. Physiol. (Lond.) 192,* 13-20.

Eccles, J.C. 1964. The physiology of synapses. Berlin : Springer.

Eccles, J.C. 1966. Cerebral synaptic mechanisms. In : Brain and Conscious Experience, 24-58, Ed. by J.C. Eccles. Berlin-Heidelberg-New York : Springer.

Eccles, J.C. 1966. Conscious experience and memory. In : Brain and Conscious Experience, 314-344, Ed. by J.C. Eccles. Berlin-Heidelberg-New York : Springer.

Elsberg, C.A. and Levy, I. 1935. Sense of smell : New and simple method of quantitative olfactometry. *Bull. NY Neural. Inst. 4,* 5.

Ekman, G. and Sjöberg, L. 1965. Scaling. *Ann. Rev. Psychol. 16,* 451-474.

Franzén, O. 1970. Neural activity in the somatic primary receiving area of the human brain and its relation to perceptual estimates. Special issue of the *IEEE Transactions on Man-Machine System. MMS-11,* 115-117.

Franzén, O. and Offenloch, K. 1969. Evoked response correlates of psychophysical magnitude estimates for tactile stimulation in man. *Exp. Brain Res. 8,* 1-18.

Granit, R. 1966. Sensory mechanisms in perception. In : Brain and Conscious Experience, 116-137. Ed. by J.C. Eccles. Berlin-Heidelberg-New York : Springer.

Hartline, H.K. 1928. *Amer. J. Physiol. 83,* 466-483.

Jones, F.N. 1958 a. Scales of subjective intensity for odors of diverse chemical nature. *Amer. J. Psychol. 71,* 305-310.

Jones, F.N. 1958 b. Subjective scales of intensity for three odors. *Amer. J. Psychol. 71,* 423-425.

Mountcastle, V.B. 1966. The neural replication of sensory events in the somatic afferent system. In : Brain and Conscious Experience, 85-115, Ed. by J.C. Eccles. Berlin-Heidelberg-New York : Springer.

Ottoson, D. 1956. Analysis of the electrical activity of the olfactory epithelium. *Acta Physiol. Scand. 35,* suppl. 122.

Reese, T.S. and Stevens, S.S. 1960. Subjective intensity of coffee odor. *Amer. J. Psychol. 73,* 424-428.

de Reuck, A.V.S. and Knight, J. (Eds.) 1966. Touch, heat and pain. A Ciba Foundation Symposium. London : Churchill. 17-26, 80-85.

Stevens, S.S. 1957. On the psychophysical law. *Psychol. Rev. 64,* 153-215.

NONVOLATILE COMPOUNDS AND THE FLAVOR OF FOODS

J. Solms
Institute of Agricultural Chemistry,
Swiss Federal Institute of Technology,
8006 Zurich, Switzerland.

Most of the current work on the nonvolatiles of foods emphasizes the importance of this fraction for nutritional purposes or as precursors for volatile flavors. Relatively little consideration has been given to the rôle of the nonvolatile fraction of foods as components of flavors (Gianturco, 1969). There are many excellent studies that have implicated tasting substances and taste effects (Amerine et al., 1965). The majority of studies, however, are concerned with the isolated compounds or with the application of these substances as intentional additives, and not with their rôle as integral components of foods. Hence it would appear that an assessment from a more empirical viewpoint, taking the composition of foods as a basis for a discussion, might lead to some interesting aspects.

By definition, tastes are limited to oral sensations of sweet, sour, salty and bitter (Moncrieff, 1967). These sensations are direct taste effects. The bitter taste is often associated with harmful compounds, exemplified by alkaloids, and is less typical for ordinary food. The salty taste is typified by sodium chloride; in many foods it is the result of added salt. Represented by sucrose and citric acid, sweet and sour tastes are produced by a variety of food constituents, which are natural components of our daily diet. In food preparations, however, the contribution of many substances exceed the direct taste effects described above, due to the phenomena of taste interactions, synergism and potentiation. These indirect effects can create important basic notes in many food flavors.

A food flavor can be formulated as a spectrum composed of several groups of compounds with close interrelationships, which give the overall flavor sensation. A tentative flavor-scheme is presented in Figure 1.

Figure 1. Composition of food flavors, tentative scheme

Flavor Sensation	Low boiling compounds and high boiling compounds with odor effects	Odor
	Potentiators and Synergists	
	Nonvolatile compounds with taste and tactile effects	Taste

The tasting substances, contributing direct taste and tactile effects, are located at the base of this model, while the volatile substances which give rise to important odor effects are situated at the top. Certain compounds can act as synergists and potentiators; they are placed in between these two fractions.

Compounds of the odor fraction are very often products of special biochemical reactions, which are linked with ripening processes, or products of chemical reactions occurring during processing of the raw material. They occur in high numbers, but in relatively low overall amounts. Compounds comprising the taste fraction are generally more closely related with the main metabolism of the raw material. They occur in much smaller numbers but generally in larger amounts in the milligram and gram range. Thus taste problems often coincide with general biochemical and nutritional aspects. The problem is best illustrated in Table 1, showing five hydrophilic model systems for foods, each composed of three main groups of constituents. All five systems are practically identical in overall chemical composition and in nutritional value. Yet their tastes vary from practically tasteless to "vegetable-like", "meat-broth like", sweet and "fruitlike". Moreover, it is easy to imagine that changes from system to system can be brought about by enzymes of the basic metabolism.

Table 1. Hydrophilic model system for foods with similar overall chemical composition and nutritional value, but different tastes.

Composition			pH	Taste
Carbohydrate (Starch)	Protein	RNA	6-7	practically tasteless
Carbohydrate (Starch)	Amino acids	RNA	6-7	"vegetable-like" taste
Carbohydrate (Starch)	Amino acids	Nucleotides	6-7	"meat-broth like" taste
Carbohydrate (Glucose)	Protein	RNA	6-7	sweet taste
Carbohydrate (Glucose)	Protein	RNA	around 3	"fruit-like" taste

Proteins and related nitrogen-containing substances - such as peptides, amino acids, nucleotides - are prominent components in many foods. Yet their rôle as taste substances has long been neglected, since they seldom exert important direct taste sensations. Isolated 5'-nucleotides have agreeable taste characteristics, which are, however, difficult to characterize (Kuninaka, 1964 a, 1964 b; Solms, 1967; Jones, 1969). Isolated L-amino acids vary in taste from flat to bitter and sweet (Solms et al., 1965). The taste of peptides has been investigated in detail only in very recent years. A few selected peptides with typical taste sensations are presented in Table 2. (Kirimura et al., 1969; Mazur et al., 1969; Virtanen, 1965, 1968).

Table 2. Selected peptides with typical taste sensations

Taste	Peptide-structure
Flat	L-Lys-L-Glu L-Phe-L-Phe Gly-Gly-Gly-Gly
Sour	L-Ala-L-Asp γ-L-Glu-L-Glu Gly-L-Asp-L-Ser-Gly
Bitter	L-Leu-L-Leu L-Arg-L-Pro L-Val-L-Val-L-Val
Sweet	L-Asp-L-Phe-OMe L-Asp-L-Met-OMe
Biting	γ-L-Glutamyl-S- (prop-l-enyl) -L-cystein

Bitter tasting peptides are often formed during enzymatic hydrolysis of proteins. Several recent publications are concerned with their isolation and structural elucidation (Murray et al., 1952; Carr et al., 1956; Harwalkar, 1967; Stone et al., 1967; Fujimaki et al., 1968; Matoba et al., 1969; Yamashita et al., 1969). Sweet tasting peptides have not been isolated from natural materials; they are products of chemical synthesis. Their possible application as non-nutritive sweeteners is now under investigation (Mazur et al., 1969). Many peptides are said to act on taste as a result of their high buffer capacity (Kirimura et al., 1969).

In food preparations, nucleotides, amino acids and peptides contribute to the flavor in a complex manner, exceeding the taste properties of the pure compounds; this is due to taste interactions, such as synergism and potentiation (Solms, 1969; Kirimura et al., 1969). The best known example is the sodium salt of L-glutamic acid (MSG), which exerts an improving effect on taste in appropriate mixtures, especially in the presence of nucleotides (Neukom, 1956; Oeda, 1963; Solms, 1967; Jones, 1969). A classical presentation of this action is the mutual effect of reduced threshold levels of glutamate in the presence of inosine-5'-nucleotide (5'-IMP) and guanosine-5'-nucleotide (5'-GMP) which is shown in Table 3

Table 3. Threshold levels of flavor potentiators

Solvent	Threshold level $^o/o$		
	5'-IMP.Na$_2$	5'-GMP.Na$_2$	MSG
(1) water	0.012	0.0035	0.03
0.1 $^o/o$ MSG	0.00010	0.000030	-
(2) 0.01 $^o/o$ IMP	-	-	0.002
(2) / (1)	1 / 120	1 / 117	1 / 15

It can be seen that the threshold levels for the recognition of glutamate and nucleotides alone are drastically reduced when these compounds occur in mixture (Kuninaka, 1964 a). Similar activities have been described for other amino acids, namely aspartic acid, homocysteic acid, ibotenic acid, tricholomic acid, *threo*-beta-oxyglutamic acid, C_7-mono-amino-dicarboxylic acids, for several peptides and for several nucleotides and nucleotide derivatives. (Kuninaka et al., 1964 b; Pfizer, 1965; Takemoto, 1966; Kirimura, 1969; Solms, 1969).

The relationship between taste activity and molecular structure has been investigated in some detail for nucleotides and nucleotide derivatives. The structural requirements for the flavor activity of nucleotides are very specific. Very few derivatives are known with increased flavor activity (Table 4). (Kuninaka, 1967; Yamaguchi, 1968; Yamazaki, 1968; Yamazaki, 1969).

Table 4. Magnitudes of the synergistic effects of various ribonucleotides and ribonucleotide derivatives with monosodium glutamate

inosine-5'-monophosphate (reference)	1.0
adenosine-5'-monophosphate	0.3
guanosine-5'-monophosphate	2.3
2-methyl-inosine-5'-monophosphate	2.3
2-ethyl-inosine-5'-monophosphate	2.3
2-N-methyl-guanosine-5'-monophosphate	2.3
inosine-1-N-oxide-5'-monophosphate	2.3
2-N-dimethyl-guanosine-5'-monophosphate	2.4
2-ethylthio-inosine-5'-monophosphate	7.5
2-methylthio-inosine-5'-monophosphate	8.0

Other compounds have more minor rôles. They do not exert typical taste stimulating activities, but seem to provide - if present in suitable mixtures - a general background effect. This effect is called ternary synergism. (Hashimoto, 1965; Yokotsuka et al., 1969; Tanaka et al., 1969 a, b). The character of the background depends on the composition of the mixture, e.g. on its content of amino acids.

The flavor contribution of the nonvolatiles of boiled potatoes has been investigated in some detail and will be discussed as an example (Buri et al., 1970). The important nonvolatiles are composed of at least three taste-contributing groups of substances, namely nucleotides, glutamic acid and other free amino acids. The composition of such a fraction of two potato varieties, namely Bintje and Ostara, is presented in Table 5.

Table 5. Free amino acids and nucleotides of boiled potatoes of the varieties Bintje and Ostara (in parenthesis) in mg per 100 g fresh material

Amino acids :

I.   Glu 73.8 (36.4)

II.   Ala 10.1 (7.4), Arg 19.8 (19.0), Asp-NH$_2$ 220.0 (187.0), Asp 46.8 (36.4) Cys-S- 1.2 (0.5), Glu-NH$_2$ 49.2 (77.6), Gly 2.2 (2.4), His 4.2 (4.3), i-Leu 10.6 (6.0), Leu 6.1 (2.9), Lys 6.8 (5.6), Met 9.2 (6.3), Phe 11.8 (4.5), Pro 9.1 (5.2), Ser 6.4 (6.4), Thr 8.0 (8.0), Try 3.0 (0.9), Tyr 11.0 (4.8), Val 25.8 (16.8).

Total amino acids : 535.1 (438.4)

Nucleotides :

III.   5'-AMP 3.0 (2.25), 5'-GMP 2.11 (1.39), 2'3'-GMP 1.72 (1.79), 5'-UMP 2.14 (1.78).

Total nucleotides : 8.96 (7.22)

An objective characterization of the taste sensation of the total nonvolatile fraction or of individual components is not possible, but an attempt can be made to determine the relative contribution of groups of substances to the taste of the overall mixture. For such an evaluation an approach described by Hashimoto (1965) for testing taste-producing substances in marine products has been adopted. It involves the stepwise reproduction of a mixture with pure compounds in water and the comparison of partially reconstituted fractions with each other in paired comparison tests with a trained taste panel, asking the judges : "Which sample is more tasty?". This approach allows a successive evaluation of the taste sensations created during the formation of a complex system. The number of "correct" answers for each step gives an approximate indication of the magnitude of the taste effects involved. The fractions as analyzed and presented in Table 5 were reconstituted with pure substances in aqueous solutions at pH 5.6. The taste properties of the fractions I, I +II and I +II +III were then compared in paired comparison tests with a taste panel. The results are presented in Table 6. The stepwise reconstitution gave a corresponding increase in taste quality for each step with results of high significance. The final total mixture had practically no odor, but an agreeable basic potato-like taste. The high significance of the results indicates that the observed taste effects were probably not simply additive, but have synergistic or potentiating character. The results obtained with amino acid fraction II were especially interesting. They confirm that free amino acids other than glutamic acid contribute with still unknown taste effects, when present in appropriate mixtures. In addition, the taste qualities of the reconstituted complete fractions from Bintje (I +II +III) and from Ostara (I +II +III) potatoes were compared with each other using the same experimental layout. In this experiment a significant preference was found for Bintje over Ostara. This result corresponds with the findings in empirical taste tests conducted with freshly boiled potatoes of the same lot and is in accordance with the general judgment on the flavor properties of both varieties.

Table 6. Results of paired comparison sensory tests of different nonvolatile fractions of Bintje potatoes (composition see Table 5)

Question : Which sample is stronger in taste ?

Pair			Ranking	Number of Judges
A	(I)	Glutamate	I < I+III	18
	(I+III)	Glutamate+ Nucleotides	I > I+III	0
B	(I+III)	Glutamate+ Nucleotides	I+III < I+II+III	18
	(I+II+III)	Glutamate+ Nucleotides+ other Amino acids	I+III > I+II+III	0

Another example reported in the literature describes the sapid components of carrot extracts, presented in Table 7 (Otsuka, 1969). The taste of carrots is said to be due mainly to the presence of glutamic acid, several sugars and the action of various amino acids around pH 5.9. Nucleotides are present only in small amounts and are therefore irrelevant to the taste.

Table 7. Nonvolatile compounds in carrot extracts

Substance	mg/100 ml *	Substance	mg/100 ml *
Aspartic acid	3.15	Tyrosine	3.74
Threonine	75.90	Phenylalanine	3.53
Serine	140.00	Tryptophane	1.60
Glutamic acid	60.50	Lysine	1.22
Proline	1.86	Histidine	1.35
Glycine	2.00	Arginine	2.73
Alanine	55.40	Taurine	20.30
Cystine	trace	Glucose	75.75
Valine	143.60	Maltose	415.90
Methionine	1.69	Sucrose	482.98
Isoleucine	5.28	Succinic acid	14.10
Leucine	3.28		

* 2O g carrot in 100 ml

Due to the fact that cheese is very rich in peptides and amino acids, which are produced by enzymatic processes during ripening, it has always been an interesting object for investigation on flavor-taste relationships (Mulder, 1952; Kosikowski, 1958; Day, 1967). The flavor of Emmentaler (or Swiss) cheese has been investigated in detail, and would appear to depend considerably on the presence of free amino acids (Langler, 1967). A typical imitation cheese flavor could be obtained when selected free fatty acids, selected volatiles and amino acids were combined and when the pH of the sample was then adjusted to that of natural cheese (around pH 5.6). Without amino acids, there was no satisfactory cheese flavor sensation.

Extracts from meat and commercial meat extract probably belong to those products best investigated. Although the volatile components are still not completely known, there are extensive data on the overall composition of these products and on possible taste effects. (Wood et al., 1957; Bender et al., 1958; Kazeniac, 1961; Wood, 1961; Batzer et al., 1962; Macy et al., 1964 a, b; Koehler et al., 1967.) An imitation commercial meat extract with interesting flavor properties is also available (Ajinomoto, 1967). Meat extracts are primarily products very rich in inosine-5′-monophosphate. Occurring in such high concentrations, it is considered to be a key basic flavor component and to be responsible for the strong meaty taste, together with other components with supporting activity (Kuninaka, 1967; Solms, 1968; Jones, 1969). The high content of inosine-5′-monophosphate is due to the presence of adenosine-5′-triphosphate as major mononucleotide in the muscle of the living animal. After slaughter there is a rapid transformation of this nucleotide to adenosine-5′-monophosphate, which is then deaminated to inosine-5′-monophosphate (Jones, 1969). A composition of meat extracts is presented in Table 8. Compounds with taste contributions reported in the literature are marked with an asterisk. Similar mixtures of specific qualitative and quantitative composition and their taste evaluations are described for fish and fish extracts (Hashimoto, 1965). There is ample evidence to suppose that amino acids, peptides and nucleotides contribute to characteristic and important basic taste sensations in the above mentioned foods.

Food commodities with vegetable character seem to be richer in glutamic and other free amino acids. Meaty products seem to rely on high amounts of nucleotides, especially of inosine-5′-monophosphate, and much less on glutamic acid. The neutrality of the taste medium seems to be important for the overall flavor sensation in the materials discussed, since it is known that glutamic acid exerts its typical taste activity only between pH 5 and 7, when it is completely dissociated. At lower pH values its taste intensity decreases rapidly (Fagerson, 1954; Heintze et al., 1958). Foods having completely different compositions are related to many fruit flavors. They contain a nonvolatile basis, composed mainly of sugars and acids, in general organic acids, with a low pH between 3 and 4.

Of all taste substances occurring in foods, sugars have long since attracted most attention, because they play equally important rôles in nutrition and in food acceptance. Intensive investigations are concerned with relative sweetness of different sugars (Amerine et al., 1965), the relationships between sweetness and molecular structure (Amerine, 1965; Shallenberger et al., 1969) and the possible replacement of sugars with non-nutritive sweeteners (Buchheim, 1965; Salant, 1968; Beck, 1969). The relative sweetness of various sugars, sweeteners and related compounds is reported in Table 9. (Buchheim, 1965; Mazur et al., 1969; Horowitz et al., 1969; Inglett et al., 1969; Beck, 1969).

Table 8. Compounds isolated from extracts from meat (Solms, 1968)

*Alanine	Anserine	Acetoin
β-Alanine	Carnitine	Acetic acid
Arginine	Carnosine	Diacetyl
Aspartic acid	*Creatine	Formic acid
Asparagine	*Creatinine	β-Hydroxybutyric acid
Citrulline	Glutathione	*Lactic acid
Cysteine	Glycerophospho-	Levulinic acid
Cystine	ethanolamine	Succinic acid
*Glutamic acid	Phosphoserine	
Glutamine	Phosphoethanolamine	
*Glycine	*Taurine	
Histidine	quat. Amines	Fructose
*Hydroxyproline	Glycoproteins	Fructose-6'-phosphate
Isoleucine	Peptides	Glucose
Leucine	Ammonia	Glucose-6'-phosphate
*Lysine	Urea	Ribose
*Methionine	*Hypoxanthine	Ribose-5'-phosphate
Methylhistidine	*Purine-Nucleosides	*Inorganic constituents
Ornithine	*Purine-Nucleotides	(mainly K-salts)
*Phenylalanine	* Inosine-5'-mono-	
Proline	phosphate and other	
Serine	nucleotides	
Threonine		
*Tyrosine	Nicotinamide-adenine-	
Valine	dinucleotide	

* Compounds with taste contributions reported in the literature are marked with an asterisk.

Table 9. Relative sweetness of various compounds

Sucrose	1	Fructose	1.1 - 1.5
Lactose	0.27	Cyclamate	30 - 80
Maltose	0.5	Glycyrrhizin	50
Sorbitol	0.5	Aspartyl-phenylalanine	100 - 200
Galactose	0.6	methylester	
Glucose	0.5 - 0.7	Stevioside	300
Mannitol	0.7	Naringin dihydrochalcone	300
Glycerol	0.8	Saccharin	500 - 700
		Neohesperidin dihydrochalcone	1000 -1500

However, there are no simple relationships between the data reported in Table 9 and the sensory action of a sweetener in foods. The relative sweetness of sugars changes with concentration, as shown for glucose and fructose in Table 11 (first horizontal and vertical columns). In mixtures of several sugars and in mixed food systems the relative sweetness of individual sugars can vary in different ways; synergistic effects with increase in sweetness as much as 20 to 30 per cent can be noted (Table 11) (Schutz et al., 1957; Pangborn et al., 1963; Stone et al., 1969 a, b). Moreover, sugars seem to contribute more than sweetness to a food; they are said to round and blend the flavor components and provide mouth-feel.

The characterization of the taste sensations of the acid components and their contribution to overall-flavors is most complex. Different acids have different tastes and the sour sensations depend on a number of factors, which include not only pH, but the total titrable acidity, buffering effects of salts and the presence of other compounds. Although many theories have been put forward, there is still much to be learned regarding the effects of acids (Braverman, 1963; Pangborn et al., 1963; Ough, 1963; Pilnik, 1964; Rudy, 1967; Moncrieff, 1967).

A collection of data for acids, illustrating the problems, are reported in Table 10 (Gardner, 1968; Wucherpfennig 1969). In mixtures,sugars and acids take part in complex interactions. The relative sweetness and the magnitude of synergistic effects are related to the acidity of the medium. Data for glucose and fructose are presented in Table 11 (Stone et al., 1969 b). Other sugars, acids and sugar-acid mixtures behave differently (Pangborn, 1963; Pangborn, 1965). Interactions were also reported with other tastes in sub- and suprathreshold concentrations (Pangborn, 1960; Kamen et al., 1961; Sjoestroem, 1955); relevant studies have been reviewed (Amerine et al., 1965). Synergistic effects with sugars have also been published for non-sugar sweeteners like cyclamate (Stone et al., 1969; Kuhr et al., 1969) and ammoniated glycyrrhizin (Muller et al., 1966). Maltol and ethyl-maltol are known to increase the flavor of sweet tasting systems without making any direct contribution in sweet taste of their own (Herrmann, 1963; Beitter, 1963; Bohnsack, 1964). In addition to these effects, interdependence of acidity, sweetness and flavor perception has been reported, especially in fruit-like beverages. (Valdes et al., 1956a, 1956b; Sjoestroem et al., 1955; Hall, 1958; Pangborn et al., 1964). All these facts suggest that the basic fruit flavor is composed of substances with close interactions.

A simple model for a fruit flavor, which is described in the literature and which is composed of a minimum of constituents, is a grapefruit flavor (MacLeod, 1966). It has the following structure :

> nootkatone (odor)
>
> sucrose   (taste)
>
> citric acid

With each compound present in right concentrations, it creates a surprisingly fruity impression and illustrates the basic elements of such a flavor. The balance between the acid and sweet components has long been recognized as very important in many natural products. As an example, data are reported for blueberries in Table 12 (Kushman et al., 1968). There is a range in composition with a medium sugar-acid ratio, which corresponds to the quality judgment "ripe".

Sugar-acid ratios were correlated with optimum quality and ripeness for many fruit preparations, giving different values for different fruit products (Amerine et al., 1965; Sinclair, 1961; Smock, 1950; Chandler, 1969). It would certainly be worthwhile to investigate what combinations of components result in the typical and different taste properties.

Table 10. Properties of various acids, arranged in decreasing order of acid taste, with tartaric acid as reference

Name	Taste in relation to tartaric acid	Total acid g / Liter	pH	Ionization Constant	Taste sensation*	Important constituent in
		measured in 0.05 N solutions				
Hydrochloric	+1.43	1.85	1.70	–	–	
Tartaric	0	3.75	2.45	1.04×10^{-3}	hard	grape
Malic	–0.43	3.35	2.65	3.9×10^{-4}	green	apple, pear, prune, cherry, apricot, grape
Phosphoric	–1.14	1.65	2.25	7.52×10^{-3}	intense	orange, grapefruit
Acetic	–1.14	3.00	2.95	1.754×10^{-5}	"vinegar"	
Lactic	–1.14	4.50	2.60	1.26×10^{-4}	sourish but tart	
Citric	–1.28	3.50	2.60	8.4×10^{-4}	fresh	berries, citrus, pineapple
Propionic	–1.85	3.70	2.90	1.336×10^{-5}	sour, mildly, cheese-like	

* Gardner, 1968

Table 11. Magnitude estimates of the relative sweetness of glucose and fructose and their mixtures in solutions of different concentrations at pH 5.8 (and 2.7 *)

(Reference of sweetness : 0.25 M sucrose at pH 5.8 = 10)

		Glucose molar concentrations				
		0	0.125	0.250	0.500	1.000
Fructose molar	0	–	0.44 (0.30)	1.07 (0.61)	4.29 (1.98)	15.73 (9.68)
concen-	0.0625	0.59 (0.28)	1.33 –	2.65 –	7.46 –	22.09 –
trations	0.125	1.68 (0.55)	2.03 (1.54)	4.14 (3.08)	10.59 (8.51)	26.88 (17.91)

* adjusted with citric acid

Table 12. Relationship between ripeness and composition for blueberries

Quality judgment	unripe	ripe	overripe
weight of berries, g	0.85	0.97	1.06
soluble solids, $^o/o$	10.1	11.7	15.0
total sugars, $^o/o$ (mainly glucose and fructose)	5.8	7.9	12;4
pH	2.83	3.91	3.76
titr. acidity, meq./100 g (mainly citric acid)	23.9	12.9	7.5
sugar to acid ratio *	3.8	9.5	25.8
amino acids, meq./100 g	1.3	1.2	1.5

* titr. acidity calculated as citric acid

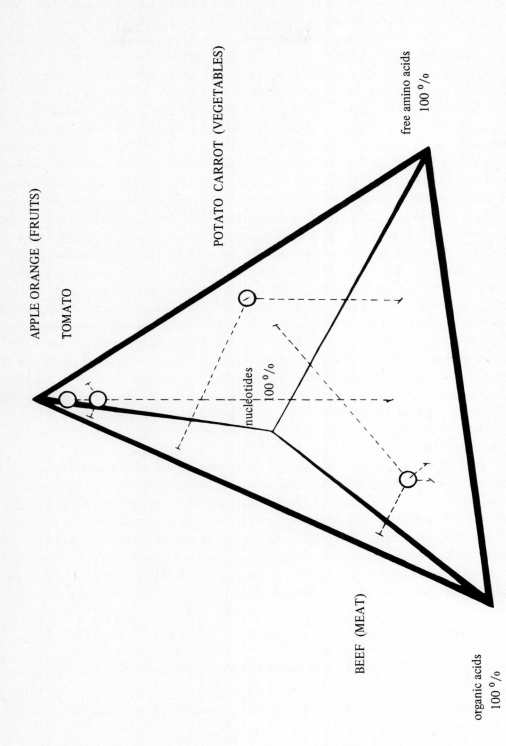

Figure 2. Four-phase graph in tetrahedral form, showing the interrelationships among amino acids, nucleotides, sugars and organic acids for different foods on a percent basis.

Table 12 indicates further that free amino acids are also present, although in smaller amounts and showing less variation. In fruit the amino acids probably have a diminished importance. The low pH of these preparations inhibits taste effects of glutamic acid, due to its incomplete dissociation, and therefore suppresses the vegetable-like taste sensations (Fagerson, 1954; Heintze et al., 1958). It is reasonable to suppose that the sugar-acid system contributes to characteristic basic taste sensations in many fruits. There is also ample evidence of taste interactions in these systems, although these effects are difficult to evaluate.

Summarizing, it is apparent that different foods have different basic tastes, related to certain groups of compounds, which exert direct and indirect taste effects. These compounds occur in varying but typical patterns, depending on the metabolic activity of the raw materials or on special methods of processing. In the foods discussed so far, amino acids and peptides, nucleotides, sugars and free organic acids have been shown to be of special importance. They are connected by various interrelationships. The existence of various patterns and effects modifying taste permits differentiation between typical taste systems. Tentative relationships between the composition of a food - that is its pattern – and its taste are visualized in figure 2. It is based on the occurrence of the four groups of taste substances mentioned above. The data were taken from the literature (Wood, 1956; Burroughs, 1957; Wood et al., 1957; Bender et al., 1958; Joslyn, 1960; Clements et al., 1962; Macy et al., 1964 b; Machado von Tschusi, 1965; Burton, 1966; Gunther et al., 1966; Solms, 1967; Satterlee et al., 1968; Grau, 1969; Otsuka et al., 1969). In a four-phase graph in tetrahedral form, the interrelationships among these four groups of substances are presented on a percent basis, calculated for beef meat, fruits and vegetables (Hoehn et al., 1970). This graph then shows a net differentiation in composition between the different foods, which parallels the differences in basic tastes. It therefore permits a general characterization of a food and its taste with a few selected parameters. Of course, such a scheme is tentative and far from complete. Several substances with interesting taste contributions, e.g. polyphenols (Horowitz et al., 1969) are lacking. One important group of food constituents, namely lipids, is not mentioned at all. Lipids exert considerable indirect taste effects due to their lipophilic properties (Mackey, 1958; Forss, 1969).

It has been an objective of this review to discuss some selected relationships between composition and taste of foods and their possible importance in the overall food flavor. Our present diet is still based on conventional agricultural products, which correspond to our flavor world. It is quite certain that in the future new agricultural derivatives and new raw materials will provide food for mankind. In order to be accepted fully, these products must be adapted to our requirements not only in nutritional value, but also in all other aspects, including odor and taste. A better understanding of our present foods in the widest sense will then be necessary for integrating these new products into our eating habits. The factor taste is one problem, the surface of which we have just started to scratch.

REFERENCES

Ajinomoto Co. Inc. 1967. Composition d'assaisonnement et produits apparentés ainsi que leur procédé de préparation. *French Pat. 1 474 613. Feb. 13.*

Amerine, M.A., Pangborn, R.M. and Roessler, E.B. 1965. Principles of sensory evaluation of food. *Academic Press, New York.*

Batzer, O.F., Santoro, A.T. and Landman, W.A. 1962. Identification of some beef flavor precursors. *J. Agr. Food Chem. 10,* 94-96.

Beck, K.M. 1969. Nonnutritive Sweeteners. In Kirk-Othmer, ed. *Encyclopedia of Chemical Technology, Ed. 2,* Wiley & Sons Inc., New York, *19,* 593-607.

Beitter, H. 1963. Ueber Maltol. *Brot und Gebäck 17,* 132-134.

Bender, A.E., Wood T.and Palgrave J.A. 1958. Analysis of tissue constituents. Extract of fresh ox muscle. *J. Sci. Food Agric. 9,* 812-817.

Bohnsack, H. 1964. Beitrag zur Kenntnis der ätherischen Oele, Riech- und Geschmackstoffe. IX. Mitt. Ueber Maltol = 2-Methylpyromekonsäure. *Riechst. Aromen Körperpflegem. 14,* 33-34.

Braverman, J.B.S. 1963. Introduction to the biochemistry of foods. Elsevier Publ. Comp., Amsterdam. p. 269-285.

Buchheim, K. 1965. Süss-Stoffe. In *Ullmanns Encyklopädie der technischen Chemie. Ed. 3.* Urban & Schwarzenberg, München., *16,* 476-485.

Buri, R., Signer V. and Solms J. 1970. Die Bedeutung von freien Aminosäuren und Nukleotiden für den Flavor von gekochten Kartoffeln. *Lebensm. Wiss. Technol. 3,* 63.

Burroughs, L.F. 1957. The amino-acids of apple juices and ciders. *J. Sci. Food Agric. 8,* 122-131.

Burton, W. G. 1966. The potato.Veenman & Zonen N.V., Wageningen.

Carr, J.W., Loughheed T.C. and Baker B.E. 1956. Studies on protein hydrolysis. IV. Further observations on the taste of enzymic protein hydrolysates. *J. Sci. Food Agric. 7,* 629-637.

Chandler, B.V. 1969. Citrus quality standards and their measurement. *Food Pres. Quart. 29,* 8-16.

Clements, R.L. and Leland H.V. 1962. An ion-exchange study of the free amino acids in the juices of six varieties of citrus. *J. Food Sci. 27,* 20-25.

Day, E.A. 1967. In H.W. Schultz, E.A. Day and L.M. Libbey, ed. *Symposium on foods : The chemistry and physiology of flavors.* The AVI Publishing Co., Inc., Westport, Conn. p. 331-361.

Fagerson, I.S. 1954. Possible relationship between the ionic species of glutamate and flavor. *J. Agr. Food Chem. 2,* 474-476.

Forss, D.A. 1969. Role of lipids in flavors. *J. Agr. Food Chem. 17,* 681-685.

Fujimaki, M., Yamashita M., Okazawa Y. and Arai S. 1968 Diffusable bitter peptides in peptic hydrolyzate of soybean protein. *Agr. Biol. Chem. 32,* 794-795.

Gardner, W.H. 1968. Acidulants in food processing. In T.E. Furia, ed. *Handbook of Food Additives.* The Chemical Rubber Co., Cleveland, Ohio, p. 247-287.

Gianturco, M. ed. 1969. Symposium on importance of nonvolatile compounds in flavor. *J. Agr. Food Chem. 17,* 677-746.

Grau, R. 1969. Fleisch und Fleischwaren. Verlag P. Parey, Berlin.

Gunther, H. and Schweiger A. 1966. Changes in the concentration of lactic acid and free sugars in post-mortem samples of beef and pork muscles. *J.Food Sci. 31,* 300-308.

Hall, R.L. 1958. Flavor study approaches at McCormick and Co. Inc. In A.D. Little Inc., ed. Flavor research and food acceptance. Reinhold, New York, p. 224-240.

Harwalkar, V.R. 1967. Comparative study of bitter flavor fraction obtained from nonbitter and bitter Cheddar cheese. *J. Dairy Sci. 50,* 956.

Hashimoto, Y. 1965. Taste-producing substances in marine products. In R. Kreuzer, ed. The Technology of Fish Utilization. Fishing News Ltd., London. p. 57-60.

Heintze, K. and Braun, F. 1958. Beziehungen zwischen der geschmacklichen Wahrnehmung von Glutamat und dem pH-Wert. *Dtsch. Lebensm. Rdsch. 54,* 25-28.

Herrmann, K. 1963. Maltol - ein Aromaförderer. *Fruchtsaft-Ind. 8,* 215-219.

Hoehn, E. and Solms, J. 1970. Unpublished.

Horowitz, R.M. and Gentili, B. 1969. Taste and structure in phenolic glycosides. *J. Agr. Food Chem. 17,* 696-700.

Inglett, G.E., Krbechek, L, Dowling B. and Wagner,R. 1969. Dihydrochalcone sweeteners - sensory and stability evaluation. *J. Food Sci. 34,* 101-103.

Jones, N.R. 1969. Meat and fish flavors, significance of ribomononucleotides and their metabolites. *J. Agr. Food Chem. 17,* 712-716.

Joslyn, M.A. 1960. Methods in food analysis applied to plant products. Academic Press Inc., New York. p. 337-377.

Kamen, J.M., Pilgrim, F.J., Gutman, N.J. and Kroll, B.J. 1961. Interactions of suprathreshold taste stimuli. *J. Exper. Psychol. 62,* 348-356.

Kazeniac, S.J. 1961. Chicken flavor. In Proceedings Flavor Chemistry Symposium. Campbell Soup Company, Camden New Jersey. p. 37-56.

Kirimura, J., Shimizu, A., Kimizuka, A., Ninomiya, T. and Katsuya, N. 1969. The contribution of peptides and amino acids to the taste of foodstuffs. *J. Agr. Food Chem. 17,* 689-695.

Koehler, H.H. and Jacobsen, M. 1967. Characteristics of chicken-flavor containing fraction extracted from raw muscle. *J. Agr. Food Chem. 15,* 707-712.

Kosikowski, F.C. and Mocquot, G. 1958. Progrès de la technologie du fromage. Etudes Agricoles, FAO, Rome. No. 38, p. 151-172.

Kuhr, W.K., Slakis, J.A., Hughes, R.L. and Neilson, J.A. 1969. Process for improving taste in fruit products by adding cyclamic acid. U.S. Pat. 3 432 305. March 11.

Kuninaka, A. 1964 a. The nucleotides, a rationale of research on flavor potentiation. In A.D. Little Inc., ed. Symposium on Flavor Potentiation, A.D. Little Inc., Cambridge, Mass. p. 4-9.

Kuninaka, A., Kibi, M. and Sakaguchi, K. 1964 b. History and development of flavor nucleotides. *Food Technol. 18,* 287-293.

Kininaka, A. 1967. In H.W. Schultz, E.A. Day and L.M. Libbey, ed. Symposium on foods : The chemistry and physiology of flavors. The AVI Publishing Co., Inc. Westport, Conn. p. 515-535.

Kushman, L.J. and Ballinger, W.E. 1968. Acid and sugar changes during ripening in Wolcott blueberries. *Am. Soc. Hort. Scie. 92,* 290-295.

Langler, J.E., Libbey, L.M. and Day, E.A. 1967. Identification and evaluation of selected compounds in Swiss cheese flavor. *J. Agr. Food Chem. 15,* 386-391.

Machado von Tschusi, E. 1965. Critères de qualité des tomates pour la consommation en frais. In Colloquium sur le potassium et la qualité des produits agricoles. Institut Internat. de la Potasse, Berne. p. 163-177.

Mackay, A. 1958. Discernment of taste substances as affected by solvent medium. *Food Research 23,* 580-583.

Mac Leod, W.D. 1966. Nootkatone, grapefruit flavor and the citrus industry. *California Citrograph 51, (3),* 120-123.

Macy, R.L., Naumann, D.H. and Bailey, M.E. 1964 a. Water-soluble flavor and odor precursors of meat. I. Qualitative study of certain amino acids, carbohydrates, non-amino acid nitrogen compounds, and phosphoric acid esters of beef, pork and lamb. *J. Food Sci. 29*, 136-141.

Macy, R.L., Naumann, H.D. and Bailey, M.E. 1964 b. Water soluble flavor and odor precursors of meat. II. Effects of heating on amino nitrogen constituents and carbohydrates in lyophilized diffusates from aqueous extracts of beef, pork and lamb. *J. Food Sci. 29*, 142-148.

Matoba, T., Nagayasu, C., Hayashi, R. and Hata, T. 1969. Bitter peptides in tryptic hydrolysate of casein. *Agr. Biol. Chem. 33*, 1662-1663.

Mazur, H.R., Schlatter, J.M. and Goldkamp, A.H. 1969. Structure-taste relationships of some dipeptides. *J. Am. Chem. Soc. 91*, 2684-2691.

Moncrieff, R.W. 1967. The chemical senses. Leonhard Hill, London.

Mulder, H. 1952. Taste and flavor forming substances in cheese, *Neth. Milk Dairy J. 6*, 157-168.

Muller, R.E. and Morris, R.J. 1966. Sucrose ammoniated glycyrrhizin sweetening agent. U.S. Pat. 3 282 706. Nov. 1.

Murray, T.K. and Baker, B.E. 1952. Protein hydrolysis. I. Preliminary observations on the taste of enzymic protein hydrolyzates. *J. Sci. Food Agric. 3*, 470-475.

Neukom, H. 1956. Ueber die technische Herstellung und die Anwendung von Natriumglutamat. *Chimia 10*, 203-208.

Oeda, H. 1963. Monosodium glutamate. In Kirk-Othmer, ed. *Encyclopedia of Chemical Technology*, *Ed. 2*, J. Wiley & Sons Inc., New York. *2*, 198-212.

Otsuka, H. and Tsuneko, T. 1969. Sapid components in Carrot. *J. Food Sci. 34*, 392-394.

Ough, C.S. 1963. Sensory examination of four organic acids added to wine. *J. Food Sci. 28*, 101-106.

Pangborn, R.M. 1960. Taste Interrelationships. *Food Res. 25*, 245-256.

Pangborn, R.M. 1963. Relative taste intensities of selected sugars and organic acids. *J. Food Sci. 28*, 726-733.

Pangborn, R.M. and Chrisp, R.B. 1964. Taste interrelationships. VI. Sucrose, sodium chloride and citric acid in canned tomato juice. *J. Food Sci. 29*, 490-498.

Pangborn, R.M. 1965. Taste interrelationships of organic acids and selected sugars. In J.M. Leitch, ed. *Food Science and Technology*. Gordon and Breach Sci. Publ., New York. *3*, 291-305.

Pfizer, Ch. & Co. 1965. Flavor enhancer and method for its production. Brit. Pat. 1 011 346. Nov. 24.

Pilnik, W. 1964. Ueber den sauren Geschmack von Fruchtsäuren. *Intern. Fruchtsaft-Union, Wiss. Techn. Kommission, Berichte 5*, 149-157.

Rudy, H. 1967. Fruchtsäuren, Wissenschaft und Technik. Dr. A. Hüthig Verlag, Heidelberg.

Salant, A. 1968. Nonnutritive Sweeteners. In T.E. Furia, ed. *Handbook of Food Additives.* The Chemical Rubber Co., Cleveland, Ohio, p. 501-563.

Satterlee, L.D. and Lillard, D.A. 1968. A procedure for gas chromatographic analysis of free amino acids in meats. *J. Food Sci. 32*, 682-685.

Schutz, H.G. and Pilgrim, F.J. 1957. Sweetness of various compounds and its measurement. *Food Res. 22*, 206-213.

Shallenberger, R.S. and Acree, T.E. 1969. Molecular structure and sweet taste. *J. Agr. Food Chem.* *17*, 701-703.

Sinclair, W.B. 1961. The orange, its biochemistry and physiology. Univ. of California, Div. Agric. Sciences, Davis, Calif. 1961, p. 156-158.

Sjöström, L.B. and Cairncross, S.E. 1955. Role of sweeteners in food flavor. *Adv. Chemistry Series 12*, 108-113.

Smock, R.M. and Neubert, A.M. 1950. Apples and apple products. Interscience Publ. Inc., New York, p. 171-172, 317-318.

Solms, J., Vuatas L. and Egli, R.H. 1965. The taste of L- and D-amino acids. *Experientia 21*, 692.

Solms, J. 1967. Geschmacksaktive Verbindungen, vor allem in Fleisch und Gemüse. In J. Solms and H. Neukom, ed. Aroma- und Geschmackstoffe in Lebensmitteln. Forster Verlag AG., Zurich. p. 199-221.

Solms, J. 1968. Geschmackstoffe und Aromastoffe des Fleisches. *Fleischwirtschaft 48*, 287-291.

Solms, J. 1969. The taste of amino acids, peptides and proteins. *J. Agr. Food Chem. 17*, 686-688.

Stone, W.K. and Naff, D.M. 1967. Increases in soluble nitrogen and bitter flavor development in cottage cheese. *J. Dairy Sci. 50*, 1497-1500.

Stone, H. and Oliver, S.M. 1969 a. Measurement of the relative sweetness of selected sweeteners and sweetener mixtures. *J. Food Sci. 34*, 215-222.

Stone, H., Oliver, S. and Kloehn, J. 1969 b. Temperature and pH effects on the relative sweetness of suprathreshold mixtures of dextrose fructose. *Perception and Psychophys. 5*, 257-260.

Takamoto, T. 1966. Studies on biologically active amino acids. *Japan Medical Gaz. 3 (5)*, 5,7,16.

Tanaka, T., Saito, N., Okuhara, A. and Yokotsuka, T. 1969 a. Taste of alfa-amino acids. II. *Nippon Nogei Kagaku Kaishi 43*, 171-176. Abs. in *Chem. Abstr. 71*, 79914.

Tanaka T., Saito, N., Okuhara, A. and Yokotsuka, T. 1969 b. Taste of alfa-amino acids. III. *Nippon Nogei Kagaku Kaishi 43*, 263-268. Abs. in *Chem. Abstr. 71*, 90095.

Valdés, R.M., Hinreiner, E.H. and Simone, M.J. 1956 a. Effect of sucrose and organic acids on apparent flavor intensity. I. Aqueous solutions. *Food Technol. 10*, 282-285.

Valdés, R.M., Simone, M.J. and Hinreiner, E.H. 1956 b. Effect of sucrose and organic acids on apparent flavor intensity. II. Fruit nectars. *Food Technol. 10*, 387-390.

Virtanen, A.I. 1965. Studies on organic sulphur compounds and other labile substances in plants. *Phytochemistry 4*, 207-228.

Virtanen, A.I. 1968. Biochemical Research Institute, Helsinki, Finland. Personal communication.

Wood, T. 1956. Some applications of paper chromatography to the examination of meat extract. *J. Sci. Food Agric. 7*, 196-200.

Wood, T. and Bender, A.E. 1957. Analysis of tissue constituents. Commercial ox-muscle extract. *Biochem. J. 67*, 366-373.

Wood, T. 1961. The browning of ox muscle extracts. *J. Sci. Food Agric. 12*, 61-69.

Wucherpfennig. K. 1969. Die Säuren - ein qualitätsbestimmender Faktor in Wein. *Deutsche Wein-Zeitg. 105*, 836-840.

Yamaguchi, S., Yoshikawa, T., Ikeda, S. and Ninomiya, T. 1968. Synergistic taste effects of some new ribonucleotide derivatives. *Agr. Biol. Chem. 32,* 797-802.

Yamashita, M., Arai, S. and Fujimaki, M. 1969. Applying proteolytic enzymes on soybean. IV. A ninhydrin-negative peptide in peptic hydrolyzate of soybean protein. *Agr. Biol. Chem. 33,* 321-330.

Yamazaki, A., Kumashiro, I. and Takenishi, T. 1968. Synthesis of 2-alkylthioinosine-5′-phosphates and N^2-methylated guanosine-5′-phosphates. *Chem. Pharm. Bull. 16,* 338-344.

Yamazaki, A., Kumashiro, I. and Takenishi, T. 1969. Synthesis of 1-N-oxides of adenosine and 2-alkyladenosine and inosine 1-N-oxide-5′-phosphate. *Chem. Pharm. Bull. 17,* 1128-1133.

Yokotsuka, T., Saito, N., Okuhara, A. and Tanaka, T. 1969. Taste of alfa-amino acids. I. *Nippon Nogei Kagaku Kaishi 43,* 165-170. Abs. in *Chem. Abstr. 71,* 79858.

DISCUSSION

Zotterman : Experiments in my laboratory demonstrated that the interaction between sugar and acids does not take place at receptor level, but higher up in the central nervous system. You mentioned that almost no proteins have taste themselves. It is known that the active principle of miracle fruit is a protein and that this protein has the property of giving a very strong sweet taste in acidic solution.

Nursten : Non-volatile constituents of fruit can have, as Dr Solms mentioned, two main effects; they can have taste or taste effects and they can be precursors of the volatiles. However, these non-volatile fruit constituents may also produce a third effect; they may interact with the volatiles, or their precursors, or the medium, and alter the balance of the system.

Solms : I concentrated only on basic factors, but not on third effects or other supplementing effects.

Kettenes : Did you ever observe that not only taste but also odor is enhanced by non-volatile substances? Did you or anybody here observe that amino acids may enhance or influence odor?

Solms : Langler et al. reported that volatiles identified from cheese produced the particular cheese flavor only on addition of amino acids.

Henning : The example of cheese flavor does not answer the question of Mr. Kettenes. Here we are dealing with a flavor and not only with an odor sensation. How do you believe flavor potentiators work? Do they influence only taste, or odor too? and should we conceive an action via the central nervous system?

Solms : I should like to give first some comments on the enhancement action of amino acids. Saliva contains all kinds of physiologically active compounds. These substances in the saliva may act with those we add. Therefore it would be very important first to know the taste-producing substances that exist in the mouth. With this knowledge we might get a better answer about interactions of added substances. I do not feel competent to discuss the physiological action of flavor potentiators, which is more the field of the physiologists present here.

Schneider :   What do you consider to be "volatile" or "non-volatile" compounds?

Solms :   We distinguish between these two groups according to the chemist's opinion on volatility. I am afraid at the present time we have no methods of high sensitivity to really measure the volatility or the non-volatility of all the substances we use. I think it would be very important for physiologists to consider this problem in order to obtain a definition of volatility and non-volatility related to odor and taste receptors.

Schneider :   Can we consider, in simple terms, that a substance acting on your nose is volatile and one acting on your tongue or mouth is non-volatile?

Solms :   In simple terms, yes.

REFERENCE

Langler, J.E., Libbey, L.M. and Day, E.A. 1967. Identification and evaluation of selected compounds in Swiss cheese flavor. *J. Agr. Food Chem. 15*, 386-391.

MOLECULAR THEORY OF ODOR WITH THE a-HELIX

AS POTENTIAL PERCEPTOR

Rudolf Em. Randebrock

D-2000 Hamburg 56, Siegrunweg 21, Germany.

THEORETICAL CONSIDERATIONS

In spite of the numerous theories of odor perception there are some facts that must fit all of them. Firstly olfaction is a chemical sense, i.e. only molecules coming into contact with the receptor sites and having the ability to interact with them can provoke an odor. It is thus necessary for all odorous substances to have a sufficient vapor pressure in order not to be completely absorbed in the nasal cavities before reaching the receptors. Secondly, the character of an odorous substance is determined by its molecular shape, while the osmophoric group only causes variations in its character (Ruzicka, 1920). Thirdly, all odorous substances have a more or less lipoid solubility. This lipoid solubility does not cause the odor sensations, but water-soluble substances seem to be more strongly absorbed on the way to the receptor sites.

Shallenberger and Acree (1967) have suggested that two hydrogen bond-forming groups at a specific distance apart in a molecule are responsible for sweet taste. This idea led to the consideration that hydrogen bonds might be responsible for odor too.

Apparently all chemically pure substances having an odor are capable of forming hydrogen bonds by means of electron-donator-acceptor complexes (Briegleb, 1961) with groups like aldehyde, ketone, ester, nitro, cyano, sulfhydryl, thioether halogens, double and triple bonds, benzene rings etc., most odorous substances having more than one such group. It is suggested that the arrangement of these secondary valency forces is responsible for the generation and the kind of odor.

If this is the case these forces must be matched by associated, equally spaced secondary valencies in the receptor, accounting for the big influence of the molecule size. Probably the receptors are built up of peptides having the capacity to form many hydrogen bonds and other secondary valencies.

A peptide acting as receptor must have a definite size, and the a-helix, besides being the construction material for many different purposes in nature, fits this criterion. Many electron micrographs show a-helices are present in the olfactory cilia, e.g. in the frog (Bloom, 1954) and the rat (Brettschneider, 1958), where they appear as in Fig. 1, an electron micrograph from the rabbit taken in our laboratory.

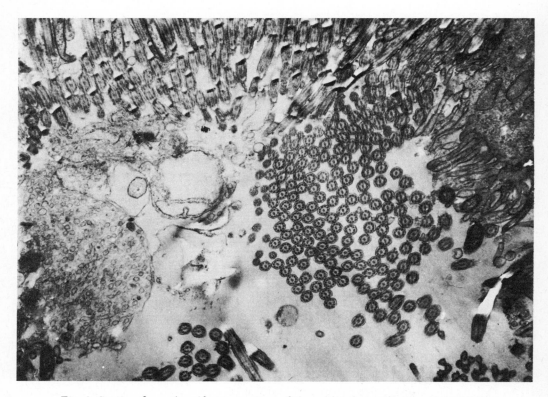

Fig. 1. Section from the olfactory region of the rabbit (magnification × 16,000).

**STRUCTURE OF THE a-HELIX, AND FUNCTION AS RECEPTOR ORGANON
(Randebrock, 1968)**

Part of the a-helix is shown in Fig. 2 (Corey and Pauling, 1955), with three hydrogen-bonded chains

$$\ldots\text{H--N--C=O}\ldots\text{H--N--C=O}\ldots\text{H--N--C=O}\ldots\text{H--N--C=O}\ldots$$

marked 1, 2 and 3, going through the whole a-helix, and having an NH group on one end, and a C=O group on the other. These three chains are largely independent of the amino acids forming

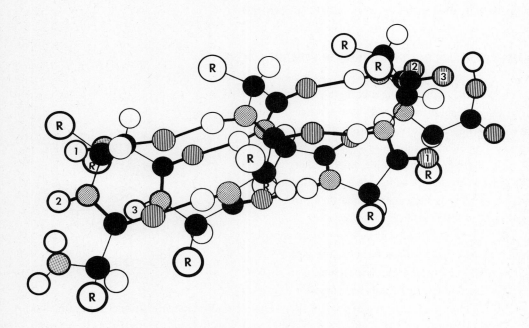

Fig. 2. a-Helix according to Corey and Pauling (1955). Partial view with the three hydrogen-bond chains.

the helix, because the influence of different amino acids only begins further back from the three hydrogen-bonded chains.

$$\ldots\text{H--N} \text{------} \text{C=O}\ldots$$

$$\text{R--CH}\qquad\text{HC--R}$$

For this reason the *a*-helix could serve as an excellent system for the transmission of energy impulses.

Thermal motion might result in a constant oscillatory frequency being formed within the three valency chains of an *a*-helix having a definite length, and if a molecule is attached to the O=C< or H–C< terminal group by means of a hydrogen bond a modulation of the oscillatory frequency could occur.

At the H–N< end of the *a*-helix are two amino acid residues, R, and the group H_2N–CH(R)–. At the O=C< site are also two amino acid residues, R, and the group HO–CO–CH (R)–. R can have about 25 different structures although some of them are too rare to be important, but there are many combinations between these residues and the H–N< or O=C< end groups, forming the different receptors, several of which may be similar in function.

If the molecule is attached not only through its hydrogen bond but also through further valencies, it can interfere longitudinally and transversely with the oscillations of the hydrogen bond chain. The model described thus leads to a two-dimensional scale for odor sensations, accounting for the wide differences between odor characteristics.

EXAMPLES OF MOLECULE-RECEPTOR COMBINATIONS

In the model of the *a*-helix the spatial arrangement of the molecular forces must have some importance in view of the importance of the molecular shape for the character of an odorous substance.

Piperonal, for example, might adhere to the H–N< group through the aldehyde group and could then fit to a tyrosin amino acid residue via a further hydrogen bond and the two benzene rings.

The macrocyclic ketones constitute another example. An isoleucine residue fits into a ring of 17-18 CH_2 units if the carbonyl group is attached to an H–N< terminal group. These compounds have an odor of civet.

Only a CH_3– group fits into a ring of 14-16 CH_2– units. The odor characteristics change correspondingly, while more than 18 CH_2– units lead to only a very weak odor (Moncrieff, 1943), when the amino acid residue can pass right through the ring. The additional CH_2– groups in ethylvanillin compared to vanillin are associated with increased van der Waals' forces explaining the increased intensity. The similarity of the odor of nitrobenzene and benzaldehyde despite their different chemical structures could be ascribed to the fact that the osmophoric groups forming the hydrogen bonds are at nearly the same distance from the benzene rings forming the second bond, and both substances have similar molecular weights.

The strong odors of molecules like diacetyl, with very low molecular weights, may be explained by reaction with two adjacent H–N< groups, the amino acid residues cushioning them against further molecular impacts. A similar explanation can be made for hydrogen sulfide and ammonia on the O–C< site.

Lower vapor pressure and higher water solubility mean less odorous substance reaches the receptor organon, so glycerol, dimethyl phthalate, etc. are odorless, despite their ample capacity for reaction with the α-helix.

There is always so much water or water vapor that normally all the $H-N<$ and $O=C<$ groups are saturated, but the H_2O molecules will be replaced by the odorous substances if their molecular forces are stronger. If insects have the same olfactory system as mammals, it might be possible for them to "smell" the water content of the air if their receptors are "dry".

EXPERIMENTS WITH THE OLFACTOMETER

If the α-helix is indeed the receptor organon, molecular models should enable prediction of whether a substance is odorous, and what sort of odor it will have. In establishing quantitatively and qualitatively odor qualities, the nose must be used as an "instrument", and the experiments discussed here consist in plotting intensity curves, extrapolation of which allows determination of the threshold value. To do this, all the odorous substances are brought to the same intensity as a comparison substance, presented in a concentration range of about 10 steps.

We found p-t-butylcyclohexyl acetate ("Oryclone") best suited as the odor for comparison. Using the olfactometer we have described elsewhere (Randebrock, 1970), this substance was diluted in the series 3×10^{-7}, 1×10^{-6}, 3×10^{-6}, 1×10^{-5} ..... 1×10^{-1}. The data for citronellal are shown in table 1, the experiments beginning   at a dilution of Oryclone at which an odor was still just perceptible. Using equation **1**, where **T** is the absolute temperature and **p** the vapor pressure, the number of molecules per c.c., **N**, was calculated for each particular dilution, **V** , the latter being determined by the settings on the apparatus (Randebrock, 1970).

$$N = \frac{6 \cdot 03p \times 10^{23}}{82 \cdot 061TV \times 10^6} \qquad \text{..................} \quad 1$$

The values for **N** at the exit of the olfactometer do not, however, agree directly with the threshold value on the nasal epithelium, and only the latter value is useful for comparison with the molecular forces. Most of the molecules are absorbed on their way through the nose, but we have no experimental knowledge of the extent of the loss for the substances we used.

We have found that substances of low vapor pressure are considerably more absorbed that those of high vapor pressure, and in order to have at least an approximation, other factors were neglected, and the concentration on the nasal epithelium was defined as :

$$N* \times \text{constant} = Np \text{ ....} 2$$

In table 1 are the experimental values for **N***, together with the corresponding logarithmic values. Table 2 shows the combined log N* values from 12 experiments for undecylenic aldehyde. The usual variation of results can be seen, not only between individuals, but also in different experiments with the same person, the largest variations being in the threshold value, the first figure in each series. Nevertheless, the average for each column was taken, and in order to have at least an order of magnitude for the deviations, σ was calculated from formula **3**, where n refers to the number of experiments.

$$\sigma = \sqrt{\frac{(\log N^*_i - \log \overline{N}^*)^2}{n(n-1)}} \quad \ldots\ldots\ldots\ldots\ldots\ldots 3$$

Usually a group of 6 experiments with the substance were put in each column. Although the deviation σ is not a measure for the absolute value of N*, this being dependent on other factors, particularly the dimensions of the individual, it nevertheless gives a good measure of comparison for the various series of experiments.

A particular dilution of Oryclone was defined as threshold value in order to eliminate the large differences between the highest dilutions in the various series of experiments. Fig. 3 shows the threshold value obtained from 287 series of experiments and expressed in concentration of Oryclone. The threshold value for Oryclone lay below dilution $V = 1 \times 10^{-6}$ in only 2.1 per cent of the cases, so the "normal" threshold level was fixed at this value.

Table 1. Data sheet from the olfactometer experiments with Citronellal.

	Citronellal		Date : Dec. 18, 1968				Person tested : K		
Dilution Oryclone	1×10^{-5}	3×10^{-5}	1×10^{-4}	3×10^{-4}	1×10^{-3}	3×10^{-3}	1×10^{-2}	3×10^{-2}	1×10^{-1}
V	318,857	58,531	43,466	5,738	3,279	1,843	895	540	338
N*	$4 \cdot 463 \times 10^6$	$2 \cdot 431 \times 10^7$	$3 \cdot 274 \times 10^7$	$2 \cdot 480 \times 10^8$	$4 \cdot 340 \times 10^8$	$7 \cdot 721 \times 10^8$	$1 \cdot 590 \times 10^9$	$2 \cdot 635 \times 10^9$	$4 \cdot 210 \times 10^9$
log N*	$6 \cdot 650$	$7 \cdot 386$	$7 \cdot 515$	$8 \cdot 395$	$8 \cdot 638$	$8 \cdot 888$	$9 \cdot 201$	$9 \cdot 421$	$9 \cdot 624$

Table 2. Log N* values of 12 data sheets for undecylenic aldehyde.

Date	Oryclone dilutions												Person tested
	3×10^{-7}	1×10^{-6}	3×10^{-6}	1×10^{-5}	3×10^{-5}	1×10^{-4}	3×10^{-4}	1×10^{-3}	3×10^{-3}	1×10^{-2}	3×10^{-2}	1×10^{-1}	
27.11.68	4·08	4·99	5·49	5·81	6·26	6·40	6·58	6·65	7·26	7·77	8·23	8·79	K
28.11.68	–	4·56	5·11	5·48	5·77	6·23	6·43	6·69	7·23	7·63	8·04	8·23	K
16.12.68	–	–	–	6·49	6·58	6·79	7·08	7·34	7·62	7·74	7·91	8·00	K
6. 1.69	–	–	–	6·23	6·74	7·11	7·42	8·04	8·28	8·73	9·20	9·63	K
6. 1.69	–	–	–	6·43	6·75	7·04	7·28	7·58	7·93	8·08	8·36	8·62	K
25.11.68	–	4·28	4·63	4·88	4·96	5·04	5·11	5·18	5·38	5·77	6·18	6·56	P
26.11.68	–	3·74	4·11	4·74	5·08	5·23	5·40	5·43	5·77	5·85	5·91	6·08	P
13.12.68	–	–	3·61	3·69	4·00	4·46	4·94	5·89	6·11	6·28	6·54	6·81	P
15.11.68	–	–	4·26	5·08	5·95	6·18	6·49	6·94	7·20	7·63	7·73	7·97	KP
14. 1.69	–	–	–	5·65	6·20	6·76	7·52	7·90	8·48	8·90	9·26	9·53	V
28. 1.69	–	–	–	5·67	6·15	6·66	7·36	7·83	8·18	8·72	9·06	9·30	V
16. 1.69	–	–	5·59	5·59	5·95	6·36	6·11	6·74	7·18	7·79	8·04	8·36	R
log N* $\varnothing$	(4·08)	(4·39)	(4·69)	5·48	5·87	6·19	6·48	6·85	7·22	7·57	7·87	8·16	
Deviation $\sigma\pm$		0·26	0·28	0·23	0·24	0·24	0·27	0·28	0·29	0·31	0·33	0·34	

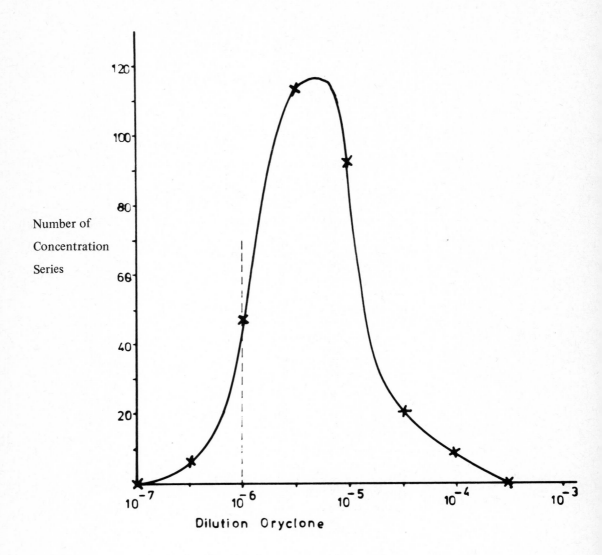

Fig. 3. Threshold values of 287 concentration series distributed according to the corresponding
        dilution of Oryclone.

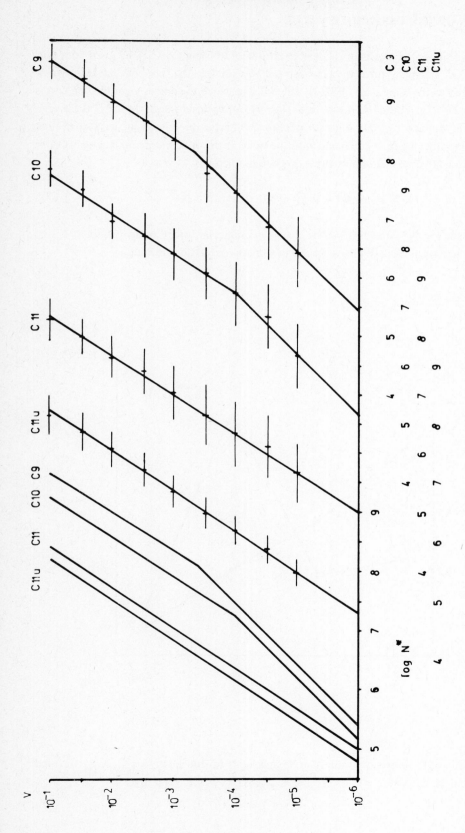

Fig. 4. Intensity curves of C_9, C_{10}, C_{11} aldehydes, and undecylenic aldehyde. The length of the lines through the measurement points gives the deviation.

CALCULATION OF THE THRESHOLD VALUE

Fig. 4 shows the results of the concentration series using C_9, C_{10}, C_{11} and undecyle-nic aldehydes. Each aldehyde shows the same increase. On the left, the four concentration series have been adjusted to the same scale of log N. The point at which the curves meet the line corresponding to $V = 10^{-6}$ gives the threshold value, log N^*, for each substance. In fig. 5, these values are compared with our approximate value for the threshold value directly on the nasal epithelium, and the threshold value, log **N**, is compared with the value at the olfactometer exit. Log **N** can be calculated from log **N*** by the equation **4** (**p** expressed in microns)

$$\log N = \log N^* - \log p + 6.0 \ldots \ldots \ldots 4$$

As fig. 5 shows, the log **N*** curve follows the change in molecular forces, unlike the log **N** curve, thus verifying – at least in this instance – the use of our approximation formula.

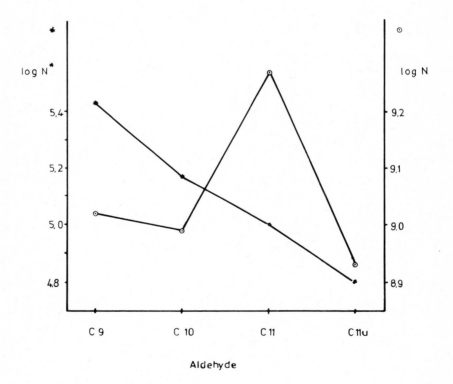

Fig. 5. Comparison of the threshold values log N* and log N for the aldehydes C_9, C_{10}, C_{11} and undecylenic aldehyde.

Table 3. Values of log N*, log p_{25^0} and log N for 12 odorous substances.
Values for p_{25^0} from Appell (1964).

	log N*	log p_{25^0}	log N
Ethylvanillin	0·58	−0·80	7·38
Vanillin	0·75	−0·64	7·39
Cinnamic alcohol	2·33	0·56	7·77
Cinnamic aldehyde	3·30	1·47	7·83
Undecylenic aldehyde	4·80	1·87	8·93
C_{11}– Aldehyde	5·00	1·73	9·27
C_{10}– Aldehyde	5·15	2·18	8·97
C_9– Aldehyde	5·43	2·42	9·11
Cineol	6·00	3·34	8·66
Nitrobenzene	6·54	2·58	9·96
Nitrosobenzene	8·84	2·91	11·93
Hydrogen sulfide	11·33	6·00	11·33

Values for log **N***, log **p**, and log **N** for all the substances examined are listed in table 3, which also shows the limits of equation 2. Highly water-soluble substances (e.g. hydrogen sulfide, nitrosobenzene) have very high values for log N* and log N, and much more of such substances is absorbed on passage through the nose with the moisture it contains. Substances comparable to the aldehydes, however, such as vanillin and ethylvanillin again show the relation with the molecular forces.

SLOPE OF THE INTENSITY CURVES

An unexpected result was found on evaluation of the intensity curves, in that the best fit of the average values gives slopes only of 0.5, 1.0, 1.5, 2.0 and 3.0 without intermediate values. Fig. 4 shows that all the aldehydes tested have the same slope of 1.5, the C_9 and C_{10} aldehydes having in addition a slope of 1.0 at lower concentrations. Fig. 6 shows that undecylenic aldehyde, cineol, and hydrogen sulfide have slopes of 1.5, 1.0 and 2.0. Clearly Oryclone must also have a continuous intensity curve, since otherwise all curves would have a discontinuity at the same point as Oryclone. Three related pairs of substances are illustrated in fig. 7. Vanillin and ethylvanillin, the latter with a somewhat lower threshold value, show the same slopes of 0.5, 1.0, and 1.5. Cinnamic alcohol and aldehyde have slopes of 0.5, 1.0 and 2.0 and 1.5 respectively, the threshold values not being comparable because of the differing water solubility. Nitrobenzene (slopes 1.0 and 1.5)

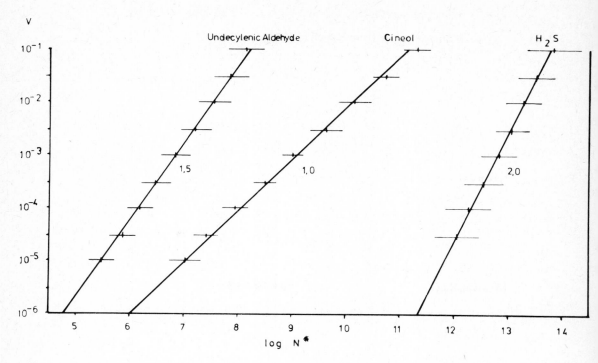

Fig. 6. Intensity slopes of undecylenic aldehyde, cineol, and H_2S.

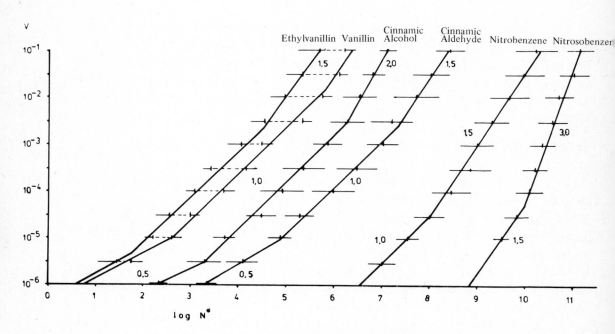

Fig. 7. Intensity curves of 6 odorants.

and nitrosobenzene (1.5 and 3.0) again have different water solubilities. Fig. 7 is typical of nearly all 40 substances we have examined, and illustrates the fact that when intensity curves are compared with Oryclone, slopes occur only in multiples of 0.5.

If we assume this to be a consequence of the law of mass action, when a molecule **M** combines with a receptor **R** to give the receptor-molecule combination **R-M**, and **z** is the number of reacting valences of the odorant,

$$[R\text{-}M] = [M]^{Z}[R]K \dots\dots\dots\dots\dots 5$$

$$\log\,[R\text{-}M] = z.\log\,[M] + \log\,[R] + \log K \dots 6$$

The compared intensity, log I, is, of course, dependent on the receptor-molecule combination, and so corresponds to log [**R-M**]. If the hypothesis is correct, **z** = 0.5 must correspond to one valence, since this is the lowest figure observed, and only multiples of it are found. Consequently, Oryclone must react with two valences over the whole length of the intensity curve measured.

The intensity curves were measured without considering the odor quality, but in many cases the individuals under test reported that the odor quality changed in the region of the discontinuity. There are probably two explanations for these discontinuities, and these are illustrated schematically in fig. 8. Intensity curve A represents an odorant reacting initially through two valences. At the discontinuity, two further valences attain threshold levels, and from there on the molecule reacts with the receptor through four valences.

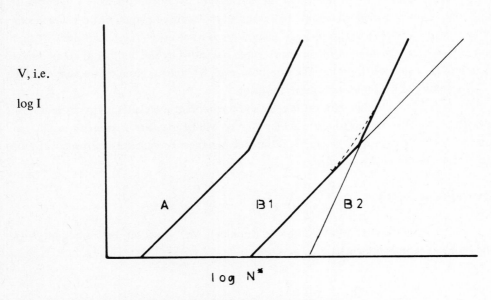

Fig. 8. Schematic description of the discontinuities in the intensity curves.

Curve B is composed of two parts. When B_2 attains its threshold level the odor quality begins to change, until at the crossover point, B_2 has the same intensity as B_1. At higher log N* values, B_2 is solely responsible for the intensity. Because the intensity, I, is given on a logarithmic scale, curves B_1 and B_2 only exert a measurable combined influence on the total intensity over the dotted part of the curve, but unfortunately the experimental results are not sufficiently exact to allow this part to be precisely determined.

According to the theory of Shah et al. (1968), delocalized π-electrons are responsible for the odor. This supports the validity of the mass action law, and our results verify some details of the theory.

CONCLUSIONS

Two types of conclusion can be drawn from our experiments. Firstly, it appears reasonable to use the human nose as an instrument for measuring odors. The receptor organ itself seems to work precisely, but it is rendered inexact by the varying state of the nostrils, that differ not only from one individual to another, but also in the same individual at different times. It is possible to reduce the error in threshold level measurement by judicious choice of the experiment, when the slopes of the intensity curves thus found are fairly exact, since they only refer to intensity changes, not absolute values. Since relative values are used, slopes, too, are reproducible over five or six concentration series, and are therefore significantly more precise than the deviation, σ, implies. These results justify a hope that reproducible results about odor quality can also be achieved.

The olfactometer experiments do not refute the hypothesis that the a-helix is the receptor organ. The results seem to confirm that intermolecular forces are responsible for the odor. The nostril is partly responsible for the fact that substances with many additional valences are odorless, since they are absorbed in the nose before reaching the receptors.

It is possible that the validity of the mass action law for the intensity curves can be explained very simply through an a-helix. In a receptor-molecule combination the osmophoric group reacts with the NH or CO group in as many ways as its electronic state will allow. If the hypothesis is correct, the slope of the intensity curve is decided by the number of possibilities of reaction of the osmophoric group. The threshold level, N*, then is given by the sum of all the forces acting between the molecule and the receptor.

The steric situation with the a-helix and the odorant molecules is known, and it is possible with molecular models to test whether the two will fit together. Interesting results can be expected particularly in cases where the difference in molecular constitution is small but differences in the odor intensity are large.

ACKNOWLEDGEMENTS

The author thanks Miss Goerke, Mr. Knobloch, Mr. Neese, Mr. Peters, and Mr. Vierk for help in the experimental work.

REFERENCES

Appel, L. 1964. Vapor Pressure. *Am. Perf. & Cosm.* 79, (5), 29-42.

Bloom, G. 1954. Studies on the olfactory epithelium of the frog and the toad with the aid of light and electron microscopy. *Z. Zellforsch.* 41, 89-100.

Brettschneider, H. 1958. Elektronenmikroskopische Untersuchungen an der Nasenschleimhaut. *Anatomischer Anzeiger, 105,* 194-204.

Briegleb, G. 1961. Elektronen-Donator-Acceptor-Komplexe. (Springer, Berlin).

Corey, R.B. and Pauling, L. 1955. *Proc. Intern. Wool Textile Res. Conf., Australia, B,* 249.

Moncrieff, R.W. 1943. *Manufacturing Chemist and Manufacturing Perfumer, 14,* 175.

Randebrock, R.E. 1968. Molecular Theory of Odor. *Nature 219,* 503-505.

Randebrock, R.E. 1970. Ueber den Einsatz eines Olfactometers. *Soc. Cosm. Chem. 21,* 289-297.

Ruzicka, L. 1920. Die Grundlagen der Geruchschemie. *Chemiker Z. 44,* 93-129.

Shah, R.K., Shaik, A.A. and Rabari, L.F. 1968. Delocalized π-Electrons and Odors *Nature 218,* 593.

Shallenberger, R.S. and Acree, T.E. 1967. Molecular Theory of Sweet Taste. *Nature 216,* 480.

DISCUSSION

Steinbrecht : What is your evidence that alpha-helical protein occurs in the olfactory receptors?

Randebrock : The a-helix is a highly important material occurring in nature in very small amounts. I have supposed that there **is** a-helical protein in the receptor, and because this was the only possibility I could think of that could build up an oscillatory system, I carried the hypothesis through to its logical conclusion.

Dravnieks : Dr. Randebrock has reported a very interesting observation with respect to the shape of the Stevens intensity slope curves. For some odorants, the plots consist of broken straight-line segments. Before attempting to develop theories for an explanation of this effect, perhaps it is worthwhile to see if such plots have more statistical significance than a curved line, or one single straight line. Dr. Randebrock has agreed to supply his original data for such a statistical test, and we hope to conduct it in the near future.

Zottermann : Is there a trigger mechanism liberating enough energy for the nervous impulse?

Randebrock : Possibly that is so.

MOLECULAR STRUCTURE AND TASTE

R.S. Shallenberger

Human nutrition research Division, Agricultural Research Service

U.S. Department of Agriculture, Beltsville, Maryland

and

Cornell University, Geneva, New York

INTRODUCTION

As knowledge of the structure of chemical compounds and general chemical theory have developed, that knowledge and theory have been periodically applied in attempts to explain why certain compounds taste sour, others salty, others sweet, and still others taste bitter. The sour taste is generally attributed to the hydrogen ion of dissociated acids. The salt taste is attributed to the anion of halide salts, with an assist from the cation. Unlike the compounds which ionize and elicit the sour and the salt tastes, compounds which elicit the sweet and bitter tastes do not ionize, and can be found in every empirical chemical class. The rationalization of sweetness and bitterness in terms of chemical theory and structure has been one of the most difficult problems of long standing. I will summarize the current state of knowledge and the most recent ideas that have been proposed to solve that problem.

CHEMICAL STRUCTURE AND SWEET TASTE

Highlights in the development of knowledge of chemical structure and chemical theory to explain sweet taste are shown in Table 1. Cohn (1914) observed that saporous groups usually occur in pairs, and that a multiplicity of hydroxyl groups usually resulted in a compound that tasted sweet. Recognizing the success of the dye chemists in relating color to the presence of an auxochrome and a chromophore in the structure of a dye stuff, Oertly and Myers (1919), proposed that in order for a compound to taste sweet, it must possess an auxogluc and a glucophore. They identified many auxoglucs and glucophores in sweet tasting compounds, but most often, the auxogluc chosen was a single hydrogen atom. Kodama (1920) pointed this out and emphasized

Table I. Historical development of theory related to compounds which cause sweet taste

Investigator	Year	Observation or theory
Cohn	1914	Saporous groups occur in pairs
Oertly and Myers	1919	The auxogluc and the glucophore
Kodama	1920	"Vibratory hydrogen"
Tsuzuki	1948	Resonance energy
Deutsch and Hansch	1966	Hydrophobicity & Hammet σ constant
Shallenberger and Acree	1967	The AH,B couple and hydrogen bonding.

that all compounds which possessed either Cohn's saporous groups, or Oertly and Myer's auxoglucs and glucophores also possessed "vibratory" hydrogen. Kodama's view of vibratory hydrogen was that compounds which elicited sweet taste existed in one or more tautomeric forms, and the transposition of the hydrogen atom to form these tautomeric forms resulted in "electronic vibration". Tsuzuki (1948) carried this idea one step further and pointed out that, in a series of closely related compounds, those with the lowest resonance energy tasted bitter, while those with the highest resonance energy tasted sweet. In 1966 Deutsch and Hansch developed an equation which correlated highly with the relative sweetness of a 2-amino-4-nitrobenzene series of compounds. They pointed out that the relative sweetness of these compounds correlated well with a "hydrophobicity" constant, π, but the best correlation arose when their equation also contained the Hammet σ constant. The importance of this last point is that the resonance energy of Tsuzuki is an index of a compounds intramolecular **electron** delocalization, and as such, is related to the Hammet σ constant. Consequently, the multiple regression equation calculated by Deutsch and Hansch correlated better with the relative sweetness of the compounds than either one alone.

In 1967, the present author, along with T.E. Acree, attempted to describe an even more general theory for the chemical characteristic common to all compounds which taste sweet. This resulted in the formulation of the AH,B theory of taste. In essence, it was proposed that all compounds which elicit the sweet taste response possess an electronegative atom A, such as oxygen or nitrogen. That atom also possesses a proton attached to it by a single covalent bond. Thus, AH represents a hydroxyl group, an imine or amine group, or in certain instances, a methine group. Within a distance of about 3 Å of the AH proton, there must reside a second electronegative atom B, again usually oxygen or nitrogen.

This general theory was an outgrowth of an intense interest in relating the structure of the sugars to their biological properties. As early as 1963 (Shallenberger, 1963), it was proposed that the reason for the varying sweetness of the sugars, when viewed as a closely related group of diasteroisomers, was due to the fact that certain vicinal OH groups could either hydrogen bond to each other, or that one of a pair of vicinal OH groups could hydrogen bond to the ring oxygen atom of the compound. When this possibility existed, it was felt that the glycol moiety, which was viewed as the repeating saporous unit in the sugars, would be restricted in its ability to elicit sweet taste. The immediate corrolary to the thesis that sugar sweetness varied inversely with the degree to which intramolecular hydrogen bonds can form was that the initial mechanism for a compound to elicit sweet taste was due to the formation of an intermolecular hydrogen bond between the saporous unit of the sweet compound and an analogous unit at the taste bud receptor site (Shallenberger, 1964).

In the continuing sugar studies, it was the recognition that sugars in a favored chair conformation yielded a glycol unit conformation with the proton of one hydroxyl group placed at a distance of about 3 Å from the oxygen atom of the neighboring hydroxyl group, and that the unit could be considered as an AH,B system conventionally used to define the hydrogen bond, that prompted us to examine other sweet tasting compounds. In essence, we believed that we could find the AH,B unit in many compounds which taste sweet, such as the sugars, saccharine, cyclamate, amino acids, chloroform, lead acetate, and the salts of beryllium. Upon publishing this idea (Shallenberger and Acree, 1967), we received a long personal communication from A.L.McClellan, who is prominent in the development of hydrogen bond theory as we know it today. He pointed out that if the thesis was correct, then an additional B moiety would be the π bonding cloud of the benzene ring, and that a compound such as benzyl alcohol should taste sweet. Of course, benzyl alcohol does taste sweet, and we had the first instance in which our thesis lead to a prediction.

Parenthetically, the suggestion that the π bonding cloud of the benzene ring could serve as a B moiety resolved the question why the *anti* isomer of anisaldehyde oxime tastes very sweet, while the *syn* isomer is tasteless. AH in the latter case is too far removed from B.

Anti-anisaldehyde oxime	*Syn*-anisaldehyde oxime
Sweet	Tasteless

    About one year after our thesis was published, some work of Warfield (1954) came to our attention. In the abstracts of the papers of a meeting of the American Chemical Society, Warfield described the unit common to all sweet tasting compounds as a "taste couple" consisting of an acidic proton and a neighboring unshared electron pair. We can only conclude that over ten years earlier, he had the same idea, and that if hydrogen bond theory had been as widely known as it is today, he would have eventually published his thoughts, and probably would have described the "taste couple" as an AH,B system.

    In relating molecular structure and the development of chemical theory to taste, it can be pointed out that the AH,B thesis contains Cohn's idea that saporous groups occur in pairs. It also encompasses Oertly and Myer's idea of auxoglucs and glucophores. The AH moiety seems to be related to Kodama's concept of "vibratory hydrogen". Substitution on a closely related series of compounds would alter the acidity and sweetness of the compounds. This is Tsuzuli's revelation described as "resonance energy". In the Deutsch and Hansch equation, AH,B accounts for the significance of the Hammet σ constant. The significance of the π constant is manifested in the fact that the sweetest sugar, fructose, also has high lipophilic character due to the methylene carbon atom.

    One final point can be made to tie the AH,B thesis into recent chemical theory. At the outset of this discussion, mention was made to the corollary to the thesis that all sweet compounds contain an AH,B system to the effect that the initial chemistry of the sweet taste response was a concerted interaction, with the formation of simultaneous hydrogen bonds between AH,B of the sweet compound and an AH,B unit at the taste bud receptor site. Deardon (1968) expressed the opinion that such a simultaneous hydrogen bond explains why the sense of sweetness is a lingering sensation, since by analogy to the carboxylic acids, such a bond would be relatively strong. Pearson (1963) developed the idea that all chemical groups can be considered to be either hard and soft acids, or hard and soft bases. Without entering into a discussion of "hardness" and "softness", mention can be made of the first principle of the theory. It is that hard acids are more prone to react with a hard base than with a soft base. The AH,B unit described contains both a hard acid moiety, AH, and a hard base moiety, B. The idea then of a simultaneous hydrogen bond between two AH,B units as the initial chemistry of the sweet taste response, fits neatly into Pearson's concept of hard and soft acids and bases.

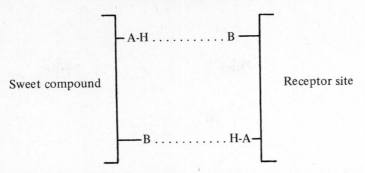

At this stage of developing the AH,B theory relating chemical structure to taste, one serious problem remained to be solved. The model shown above to describe the initial chemistry of sweet taste is two dimensional, but important data are available which demand the inclusion of a third dimension in the model. Foremost among these data is the fact that many amino acids which belong to the synthetic D-series taste sweet while their mirror-image compounds (the L-series) are tasteless. However, the simplest amino acid, glycine, and the next member of the homologous series, either D- or L-alanine taste sweet. Since the anomalous taste of the amino acids arose only when the amino acid chain substituent was larger than the ethyl group, we (Shallenberger et al., 1969) proposed that the third dimension for the taste bud receptor site need only be a spatial barrier. As shown in Fig. 1, an amino acid has a fixed AH,B unit, and can make only one approach to the receptor site. If the side chain is larger than the ethyl group, as it is in leucine, as an example, then a spatial barrier at a distance of about 3-4 Å from the AH,B site serves to explain why the enantiomorphic amino acids yield different tastes.

Because they are biochemically inert, and through analogy to the tastelessness of the L-amino acids, it has been widely assumed that the L-series of sugars are tasteless. Indeed, there is one report in the literature to the effect that L-glucose is tasteless. Since vicinal hydroxyl groups are postulated to be the AH,B unit of the sugars, and since a single hydroxyl group can act as either AH or B, the taste model site proposed seemed inadequate. A sugar, whether D- or L- can make a variety of approaches to the site and hence, the proposed model requires that the L-series of sugars should be at least as sweet as the D-series. Through the courtesy of Dr. N. K. Richtmyer, we were able to obtain enantiomorphic forms of the sugars glucose, galactose, mannose, arabinose, xylose, rhamnose, and glucoheptulose. Upon submitting these pairs of sugars to a taste panel, we were excited to find that the panel could not distinguish between the sweet taste of the enantiomorphic sugars. In confirmation of our findings, Wolfrom and Thompson (1948) reported that L-fructose tastes "very sweet". Thus, the notion that L-sugars are tasteless appears to be only myth, and the taste model site proposed can account for the taste of enantiomorphic amino acids and sugars. Furthermore, the initial chemistry of the sweet taste response appears to be physicochemical in nature, as required by the AH,B thesis, and not enzymatic in nature.

CHEMICAL STRUCTURE AND BITTER TASTE

Very little information is available which would permit one to deduce the chemical grouping common to those compounds which elicit the bitter taste response. We have concluded (Shallenberger and Acree, 1970) that there must be more than one site, and that at least one of these sites must be tripartite in a strict chemical sense, i.e., the compound has three reactive sites. As for the other sites, the recent work of Kubo and Kubota (1969) is interesting. Those investigators report that in a series of diterpenes isolated from plants, the common group for bitter taste is the AH,B unit needed for sweet taste, but that the AH proton to the B orbital distance is 1.5 Å ,

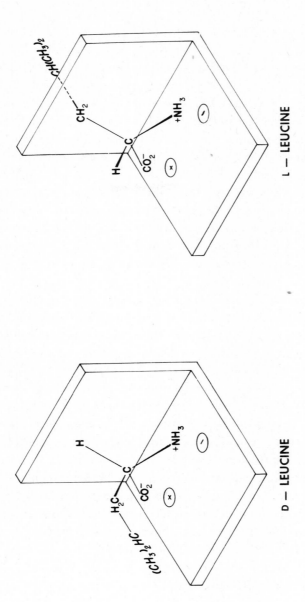

Fig. 1   Reaction of a D-amino acid at the AH,B receptor site to elicit sweet taste, and the inability of the enantiomorphic form to be positioned over the site resulting in tastelessness. From R.S. Shallenberger et al., *Nature 221*, 555 (1969) and reproduced by permission of the publisher.

resulting in a strong intramolecular hydrogen bond. In other words, while the sweetness of the sugars varies inversely with the formation of hydrogen bonds, the intramolecular hydrogen bond is a requisite for bitterness in the series of diterpenes studies.

In our early work (Shallenberger, 1963), one of the key comparisons that lead to the idea that sugar sweetness varied with intramolecular hydrogen bonding was consideration of the structure of glucose and its diasterisomer galactose. The only difference between these two compounds is that the hydroxyl group on carbon atom number four is equatorially disposed for glucose but the hydroxyl group is axially disposed in galactose. The steric disposition of the axial hydroxyl group in galactose is such that it can form an intramolecular hydrogen bond with the ring oxygen atom. This was offered as the reason why galactose is only about one-half as sweet as glucose. One glycol group of the sugar galactose would be restricted in its ability to form an intermolecular hydrogen bond with the taste bud receptor site.

Glucose   Galactose

Recently I have had the good fortune to be able to work closely with Dr. Robert I. Henkin at the National Institutes of Health at Bethesda, Maryland. Dr. Henkin was treating two patients for hypoparathyroidism who also had a very unusual sensory defect. While their taste sense was normal with respect to sour, salty, and bitter substances, they could not recognize sweetness. We (Henkin and Shallenberger, 1970) have named this phenomenon aglycogeusia (Gr. ageusia, without taste; glyks, sweet).

When saturated sugar solutions were submitted to these patients, their response offered valuable information on the structure of these compounds. Fructose, the sweetest sugar also possesses the most acidic anomeric hydroxyl proton. In the absence of an ability to detect sweetness, the patients consistently described solutions of fructose as tasting sour. In view of the theses that sugar sweetness varies inversely with intramolecular hydrogen bonding, and that intramolecular hydrogen bonding is a requirement for bitterness in certain compounds, we found it interesting that the patients consistently described solutions of galactose as tasting bitter. Thus, in the absence of ability to detect sweetness, the patients responded to galactose in a manner which requires an intramolecular hydrogen bond.

In closing this discussion on the relation between molecular structure and taste, the material presented can be summarized in the following manner. The initial chemistry of sweet taste appears to be a concerted simultaneous interaction between an AH,B unit of a saporous compound, and an analogous AH,B unit at the taste bud receptor site. The forces involved appear to be the formation of intermolecular hydrogen bonds. In one instance, the bitter taste appears to be an interaction at the same site, but because the tastant is strongly hydrogen bonded intramolecularly, the interaction may be non bonded, or repulsive. The reaction of a dissociated proton at the negative dipole of the site through ionic forces may be the initial chemical basis of the sour taste response, while the reaction of a halide anion at the positive dipole (AH) moiety of the site through ionic forces may be the initial chemical interaction for the salt taste response. If this is not the case, and there are various sites for the different senses of taste, they may have those general chemical characteristics just described.

ACKNOWLEDGMENT

Many persons have had a keen interest in the material just presented since an attempt is being made to relate chemical structure to sensory response. Indeed, many valuable suggestions and ideas presented are not just those of the author, but of one or more persons. Some of them are Professors T.E. Acree, C.Y. Lee, Ada and Professor John Hill, Professors J.S. Brimacomb, J.K.N.Jones, L.M. Beidler, S.Price, R.U. Lemeuix, and Dr. N.K. Richtmyer.

REFERENCES

Cohn, G. 1914. Die Organischen Geschmackstoffe. *Siemenroth,* Berlin.

Deardon, J.C. 1968. The hydrogen bond. *New Scientist 37,*628-630.

Deutsch, E.W. and Hansch, C. 1966. Dependence of relative sweetness on hydrophobic bonding. *Nature 211,* 75.

Henkin, R.I. 1970. Aglycogeusia : The inability to detect sweetness and its possible molecular basis. *Nature, in press.*

Kodama, S. 1920. Taste. *J. Tokyo Chem. Soc. 4l,* 495-534.

Kubota, T. and Kubo, I. 1969. Bitterness and chemical structure. *Nature 223,* 97-99.

Oertly, E. and Myers, R.G. 1919. A new theory relating constitution to taste. *J. Amer. Chem. Soc. XLI,* 855-867.

Pearson, R.G. 1963. Hard and soft acids and bases. *J. Amer. Chem. Soc. 85,* 3533-3539.

Shallenberger, R.S. 1963. Hydrogen bonding and the varying sweetness of the sugars. *J. Food Science 28,* 584-589.

Shallenberger, R.S. 1964. Why do sugars taste sweet? *New Scientist 23,* 569 .

Shallenberger, R.S. and Acree, T.E. 1967. Molecular theory of sweet taste. *Nature 216,* 480-482.

Shallenberger, R.S., Acree, T.E. and Lee, C.Y. 1969. The sweet taste of D- and L-sugars and amino acids and the steric nature of their receptor site. *Nature 22l,* 555-556.

Shallenberger, R.S. and Acree, T.E. 1970. Chemical structure of compounds and their sweet and bitter taste. In *Handbook of sensory physiology IV, Chemical Senses* (L.M. Beidler, Editor) Springer-Verlag, Berlin, in press.

Tsuzuki, Y. 1948. Sweet taste and chemical constitution, *Chemistry and Chem. Ind. (Japan) 1,* 32-40.

Warfield, R.B. 1954. Taste and molecular structure. *Abstr. Papers Amer. Chem. Soc. 126th meeting, Div. Agr. and Food Chem., New York, N.Y.,* p. 15a.

Wolfrom, M.L. and Thompson, A. 1948. L-Fructose. *J. Amer. Chem. Soc. 68,* 791-793.

DISCUSSION

Le Magnen :   It is known that in different insect species, and in different vertebrates too, the order of activity of the various sugars is not the same. How do you explain these differences?

Shallenberger :   I do not know. This theory does not have much bearing on organisms other than human.

Beets :   In the case of the aromatic oximes you mentioned, where I see no conceivable way how an unactived benzene ring can be an effective hydrogen donor, nor can I see the aromatic system as a π-electron donor. Would it not be more probable that the AHB pair is really located in the oxime group and not in the benzene ring and that it is merely a matter of the benzene ring hindering or not hindering the interaction?
Your diagram of L-leucine has the isopropyl group sticking out towards the back of the picture; we should not forget that isopropyl groups can rotate freely and can also be drawn in front of that barrier so that the same effect should be shown by alanine for instance, where there is only a CH_3 group. Finally I wonder whether the bifunctional polar group is not a general phenomenon in all tastes in contrast to odor, where we have generally to do with a monofunctional group.

Shallenberger :   Yes, it could be the functional group for all tastes. We have drawn leucine using conformational principles, which suggest that the one drawn would be the favored structure. As to your question concerning the oximes, I agree, I think we may have picked the wrong way to look at the molecule and for the acidic proton, but the π-bonding cloud of the benzene ring is certainly an electronegative center.

Shallenberger :   (in reply to an inaudible question).
I have been asked to comment on the sweet taste of certain heavy metal salts. Lead acetate ("sugar of lead" in the old literature) and beryllium chloride ("glucinium" = sweet) are reported to taste sweet. These salts do not dissociate in the usual manner, but form hydrates. Through conductivity studies, it was established that beryllium chloride in water forms a 6-membered ring containing 3 beryllium atoms and 3 water molecules, and models show a similarity of this ring with the favored chair conformation of the hexose sugars, and so within the hydrated cyclic beryllium complex, there is an AHB system in the vicinal hydroxyl groups of molecules of water of hydration.

MOLECULAR MECHANISMS OF OLFACTORY DISCRIMINATION

AND SENSITIVITY

R.H. Wright and Ronald E. Burgess

B.C. Research, Vancouver and University of British Columbia, Canada

INTRODUCTION

The scientific study of olfaction calls for answers to three questions : How is discrimination effected? What determines stimulus specificity? How is sensitivity achieved?

The ability of the vertebrate nose to discriminate between many thousands and perhaps millions of odor sensations calls for a very high information-handling capacity in the olfactory apparatus. This has been most plausibly interpreted on the basis of qualitatively specific sensitivities in the receptor nerve cells (Hainer et al. 1954). Such qualitative specificities were subsequently demonstrated in the frog (Gesteland et al. 1963). With significant empirical correlations between low-frequency molecular vibration patterns and the specific odors of bitter almond, cumin, musk, and very recently the "green" odor, and with the successful prediction of insect-attracting or odorous properties in previously untested compounds on the basis of purely spectroscopic evidence, a foundation has been laid for an understanding of stimulus specificity (Wright and Robson, 1969; Wright and Burgess, 1969; Wright, 1970). Finally, if the initial process of receptor nerve excitation is one not involving any "capture" or other permanent alteration of the stimulus molecules, sensitivity is less of a problem because each molecule can excite a large number of receptors during its passage over the receptor surface.

Thus the picture of olfactory perception that emerges involves complex sensations made up of a plurality of **primary olfactory responses** (olfactory =in the nervous system of the organism) so that a primary olfactory sensation would result from the stimulation of one type of receptor cell only. Each kind of primary olfactory response is generated by a **primary osmic stimulus** (osmic = in the odorous molecule). Since each molecular species may and usually does carry several primary osmic qualities, it follows that few if any olfactory sensations are primary (Wright, 1966).

The primary osmic qualities of molecules appear to be intimately related to their low-frequency vibrational movements ("normal modes") but additional factors such as the mechanical character or symmetry of the molecule and the oscillation may govern their osmic activity or significance.

EMPIRICAL VIBRATIONAL CORRELATIONS

The suggestion that odorous character is related to vibrational specificity is a very old one, which did not become capable of detailed experimental investigation until the application of thermodynamic considerations appeared to set an upper limit to the frequencies likely to be significant at approximately 500 cm^{-1} (Wright, 1954; Wright and Burgess, 1970). It now appears that the upper limit of effective frequencies may be as high as 1100 cm^{-1}. Whether there is a low frequency limitation is still undetermined.

Originally what was sought was an association of particular olfactory sensations or insect-attracting properties with particular bands or combinations of bands in the far infrared absorption spectra of molecules. More recently, as a larger body of spectroscopic data has become available, it appears that a particular olfactory response may depend upon the absence of some frequencies as well as upon the presence of others. For example, a study of compounds known to attract "yellow jacket" wasps *(Vespula* spp.) showed a significant correlation with the simultaneous presence of bands near 340 and 400 cm^{-1} and with an additional requirement that bands near 420 and below about 225 cm^{-1} should be weak or absent. On the basis of these spectroscopic criteria some very attractive wasp-attractants were discovered (Wright, 1970).

Again, a very recent study of a number of perfume chemicals generally regarded as possessing a distinctively "green" character indicated a positive association with a band near 225 cm^{-1} together with an additional band or bands near 285 and 310 cm^{-1}, and with the absence of any band near 260 cm^{-1}. Using these specifications a file of approximately 500 spectra was searched and fifteen substances were selected. In the opinion of several expert perfumers, at least six of the fifteen were judged to possess the "green" character.

Yet again, whereas an earlier study of the far infrared spectra of musk perfumes showed a positive correlation with four bands near 160, 250, 310 and 405 cm^{-1} (Wright and Burgess, 1969), an extension of the study to include a total of 64 "musks" has revealed an even more striking absence of peaks near 190, 225 and 265 cm^{-1}. This is shown in Fig. 1 which gives the "peak density" or number of compounds having an absorption maximum in a band 7 cm^{-1} wide as the band is moved across the spectral range from 100 to 500 cm^{-1}. What is most striking is the fact that whereas 23 of the 64 musks exhibit a peak in the range 159-165 cm^{-1}, **not one** of the 64 has a peak between 261 and 267, and only one has a peak between 189 and 195 cm^{-1}.

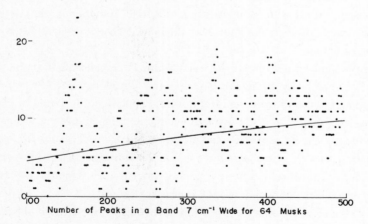

Fig. 1. Peak density plot for 64 chemicals with a musk-like odor
showing the deviation from the Ideal Control Population
indicated by the solid line.

The same thing is shown even more clearly by the data for 59 compounds reported by Beroza and Green (1963) to be attractive to the Oriental fruit fly, *Dacus dorsalis* Hendel, shown in the peak density plot in Fig. 2. In place of a positive correlation of the attractant property with a single frequency such as the one near 200 cm^{-1}, we have a plot that undulates up and down and which approaches zero in 7 cm^{-1} wide bands centering on 186, 215, 250, 364 and 417 cm^{-1}.

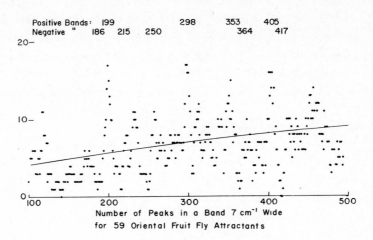

Fig. 2. Peak density plot for 59 substances which attract the Oriental fruit fly, showing the deviation from the Ideal Control Population indicated by the solid line. Frequencies of "positive" and "negative" bands are shown at the top.

By averaging the peak density for more than 500 miscellaneous organic compounds, we find that the probability of a peak appearing in a band 7 cm^{-1} wide increases as the square root of the frequency and is given approximately by $0.007\,\nu^{1/2}$. The solid curves in Fig. 1 and 2 represent this average or "Ideal Control Population" for random collections of 64 and 59 compounds respectively. It is significant that the peak density plot for 187 compounds rated simply as "non-attractant" to the Oriental fruit fly does not show the same degree of departure from the norm as does the group of attractants.

It is appropriate here to examine the real meaning of certain statistical criteria of significance. In Fig. 2, 17 out of 59 fly attractants have an infrared peak between 196 and 202 cm^{-1} whereas in 59 compounds drawn from the Ideal Control Population this should be expected for no more than 6. Thus whereas 29 % of the attractive compounds have a peak in the chosen band, only 10 % of randomly chosen control compounds would have a peak there. This gives a parameter of association of 0.19 with a chi-square test value of 5.4 indicating that the association is significant at the 0.02 level. Thus there is a positive partial association between the 199 ± 3 cm^{-1} band and attractancy which although it is weak is nevertheless real.

In view of the more abundant information now available, the original objective which was to associate particular olfactory responses with the presence of certain combinations of bands must be modified to include combinations in which certain bands are systematically absent. This does not entail any radical revision of the vibrational hypothesis : all it implies is that in the course of synaptic processing of the received information, a positive behavioral response by an insect may be inhibited or that, in the case of a vertebrate the quality of the sensation may be altered as much by the absence as by the presence of certain bands. Indeed, the effect may sometimes be generated at the level of the primary receptors where it has been shown that the discharge of single olfactory nerve fibres in the frog may be activated by some substances and inhibited by others. (Gesteland et al. 1965). Similar results have been reported for insect receptors by Boeckh (1967).

STATISTICAL ASPECTS OF INSECT ATTRACTION

If ν is taken from 100 to 500 cm^{-1}, the number P (ν) of infrared absorption peaks in a randomly selected group of M compounds in a narrow wave number interval $\delta\nu$ has the mean value

$$\overline{P}(\nu) = 10^{-3} M \nu^{1/2} \delta\nu$$

in the vicinity of frequency ν. This represents the Ideal Control Population mentioned above. For a randomly chosen set of M compounds, the peak density follows this behavior with only statistical fluctuations around the average. Since the statistics of the number of peaks is close to Poisson, the standard deviation in P(ν) about $\overline{P}(\nu)$ is $[\overline{P}(\nu)]^{1/2}$. In a typical situation with M $=60$, $\nu = 225$ cm^{-1} and $\delta\nu$ chosen as 7 cm^{-1}, the value of $\overline{P}$ is 6.3. The sharp maxima and minima above and below the average shown in Fig. 2 display the association between the infrared absorption spectra of the compounds and their ability to attract the Oriental fruit fly.

The appearance of deep minima or nulls is striking because they show that all or very nearly all the attracting compounds have no infrared peaks at these frequencies, and from this it may be inferred that peaks in these "negative bands" must in fact be absent from the spectrum of a compound if it is to elicit the response. The probability of a null arising by statistical fluctuation is approximately

$$.\exp\left[-\overline{P}(\nu)\right]$$

which is 0.0018 for $\overline{P} = 6.3$.

The maximum values in the peak count are always found to be less than the number of compounds, for example, 17 compared with 60. The probability for such a value to arise by chance is 0.0002. It is therefore inferred that no one infrared frequency is mandatory in a compound for it to elicit the response, but instead some sub-set of the positive bands suffices. The Oriental fruit fly appears to represent a simple case in which an attractant need have a peak in only one of the positive bands.

The association of attractiveness to the Oriental fruit fly with a pattern comprised of several vibrational frequencies, some "negative" and some "positive" in their effect makes it necessary to reconsider the implications of the large-scale testing program of the U.S.D.A. The results were previously interpreted as suggesting that insects with a response probability in the vicinity of 25% were responding to the stimulation of a single primary, and those with lower response probabilities, such as the Mexican fruit fly and the gypsy moth, responded only to the simultaneous stimulation of more than one kind of primary receptor (Wright, 1966). It now appears that the high response probability of the Oriental fruit fly is associated with a positive response to any one of several positive stimulus bands and it is necessary to enquire whether the observed response probability is compatible with the kind of multi-primary mechanism now envisaged.

For this insect, the U.S.D.A. found 775 attractants out of 3281 compounds tested, giving a fraction of 0.236. The population of the compounds tested may be regarded as essentially random since it included hydrocarbons, acids and anhydrides, aldehydes and acetals, esters and lactones, ethers, ketones, alcohols and phenols, and compounds containing nitrogen, or halogen, or sulfur, or phosphorus, and attractants were found in every class.

Fig. 2 suggests that attractancy is associated with the presence of any one of four positive bands at 199, 298, 353 and 405 cm^{-1} and with the simultaneous absence of all five negative bands

at 186, 215, 250, 364 and 417 cm^{-1}. Taking the frequency interval $\delta\nu = 7$ cm^{-1} used in preparing the plot as the band width of the receptors associated with these bands, the probabilities that
a randomly chosen compound will have a peak in the four positive positions are respectively 0.099,
0.121, 0.131, and 0.141 so that the probability that the random compound has a peak in one or
more of these bands is

$$1 - (1 - 0.099)(1 - 0.121)(1 - 0.131)(1 - 0.141) = 0.41$$

The probabilities that a random compound has peaks in the negative bands are respectively 0.095,
0.103, 0.110, 0.134 and 0.143. Thus the probability of no peak appearing in any of the negative
bands is

$$(1 - 0.095)(1 - 0.103)(1 - 0.110)(1 - 0.134)(1 - 0.143) = 0.54$$

Finally, the probability that a randomly chosen compound complies with both parts of the specification for Oriental fruit fly attraction is

$$0.41 \times 0.54 = 0.221$$

which is to be compared with the value 0.236 from the U.S.D.A. testing program.

This agreement shows that the elaborate band model with 4 positive bands and 5 negative
bands can still lead to an appreciable fraction (approximately $1/4$) of randomly chosen compounds
being attractants. The biological advantage of such an elaborate specification is to enable the insect
to avoid inappropriate stimuli as well as to respond to those which natural selection has established
as advantageous.

In the statistical analysis of the U.S.D.A. data on insect attractants (Wright, 1966) it was
pointed out that if F_1 is the fraction of the large population of compounds which attract insect
No. 1, and F_2 is the fraction for insect No. 2, then the fraction F_{12} which attracts both insects can
obey any of the three relations

$$F_{12} \gtreqless F_1 F_2$$

The interpretation of the equality $F_{12} = F_1 F_2$, which is the case that occurs most commonly, was that the two insects were responding to different and independent osmic qualities
within the same molecule. It now appears to include the further implication that the receptor specifications (band patterns) for the two insects are wholly independent also.

In a minority of insect pairs, The U.S.D.A. data show that $F_{12} > F_1 F_2$ which implies
some common feature, or "overlap" in the band patterns for the two insects.

From the seven insect species comprised in the U.S.D.A. survey, 21 pairs can be selected
and in one pair only was there a marginally significant indication that $F_{12} < F_1 F_2$. This was for-
merly interpreted as signifying that the presence of one "primary osmic quality" in a molecule in
no way precluded the presence of any other. If the existence of negative bands is admitted, the interpretation may be different : that in the seven insects concerned, with one possible exception, a
positive band for one insect does not coincide with a negative band for another.

MOLECULAR BASIS OF RECEPTOR SPECIFICITY

A consideration of the way in which the characteristic vibrational spectrum of the stimulus could be conveyed to the organism assumed that the interaction between the stimulus molecules and the receptor involved an energy transfer by virtue of dynamic dipole-dipole interaction at
distances of the order of 1.5 to 3 Å (Wright and Burgess, 1970). At that time it was supposed that
the interaction took place within the aqueous layer covering the receptor surface, but it now seems
more likely that the stimulus molecules may act while still in the vapor phase. The act of inhala-

tion results in high air speeds over the receptors with Reynolds' Numbers in excess of 1000, and the effect of such turbulent flow is to maximize the opportunity for any odorous molecules to approach the surface at the same time that it churns up the aqueous layer in which the olfactory cilia float and thereby bring them into contact with the air.

Vibrational energy exchange between unlike molecules is known to be strongest for near-resonant interaction and a number of such reactions (for example, CO_2 and N_2) have been studied in detail. If there is exact resonance, the probability of energy transfer during a collision is inversely proportional to the relative velocity. For a near-resonant condition, there is an optimum velocity for which energy transfer is most probable and this velocity is greater the larger the energy mismatch. Since the interaction occurs at distances very much smaller than the wave length of the vibrational mode ($1/\nu \approx 10^{-2}$ cm) there is no retardation and the transition probability assumes a simple form based on electrostatic considerations.

Energy transfers between gas molecules of a single kind and also between molecules of different kinds but in which the transition energies approximately coincide have been reviewed by Cottrell and McCoubrey (1961) and Stevens (1967) while a simple model for near-resonant energy transfer has been considered by Yardley (1969). The results are fully consistent with a process of primary receptor stimulation in which specificity of response is determined by vibrational resonance.

It is sometimes objected that this mechanism is ruled out on thermodynamic grounds because at any given temperature the population of each vibrational state is rigorously fixed by the temperature, from which it follows that as long as the temperature remains uniform the addition of new kinds of molecules can not alter the distribution of energies. However, even when the temperature is everywhere uniform, a living biological system is not in thermodynamic equilibrium and the population of vibrational levels in the molecules of the receptor cell would not be that corresponding to the prevailing temperature but may, instead, be that corresponding to a higher temperature. The effect of a species of metabolic "pumping" may be to alter and perhaps even invert the normal distribution of vibrational energies in the receptor in which case the role of the stimulus molecules would be to effect a highly specific de-excitation of the system which could be registered as a significant event by the organism.

POSSIBLE ROLE OF WATER MOLECULES

Several investigators have found that the probability for vibrational energy transfer in a gas can be greatly increased by the presence of a foreign gas (see Cottrell and McCoubrey, 1961). It is particularly striking that of the foreign molecules, water is among the most efficient in deactivating vibrationally excited gases. The increase is typically of the order of $10^2 - 10^4$ as compared with collisions between molecules of the host gas. Molecules of the foreign gas usually have a set of rotational or vibrational frequencies which permit correspondence with those of the host molecules and so facilitate the energy transfer.

Since water vapor is thus an efficient energy transfer catalyst and is present at a concentration of about 10^{18} mols/cc in the nasal cavity, there is a possibility that it plays a part in the exchange of vibrational energy between the stimulus and the receptor. At 310K only a negligible fraction of the water molecules would be excited to the first vibrational level (1595 cm^{-1}) but the numerous rotational states up to about 300 cm^{-1} will be populated with a high degree of probability. These rotational states and the transitions between them have been extensively studied and the absorption peaks in the range $0.5 - 500$ cm^{-1} are well known. Some of these rotational transitions may act as intermediaries to provide conditions for vibrational energy matching that might not otherwise occur. This should be manifest in the peak density plots of Fig. 1 and 2, and especially if these could be extended to frequencies below 100 cm^{-1}.

A variety of reactions between the stimulus molecules, the water molecules and the receptor can be considered :

$$\text{(a)} \quad S + W + R^* \; \longleftrightarrow \; S^* + W^* + R \qquad \nu_s + \nu_w = \nu_r$$

$$\text{(b)} \quad S + W^* + R^* \longleftrightarrow \; S^* + W + R \qquad \nu_s - \nu_w = \nu_r$$

$$\text{(c)} \quad S^* + W + R^* \longleftrightarrow \; S + W^* + R \qquad \nu_s - \nu_w = -\nu_r$$

The double arrow implies that the frequency relation given on the right is applicable to either direction of the reaction; the forward reaction applies to de-excitation of the receptor whereas the reaction to the left corresponds to excitation of the receptor by the stimulus.

Since the available evidence for the vibrational model of olfaction calls for values of ν_s up to about 500 cm^{-1} while the intense rotational lines for H_2O involve values of ν_w up to 300 cm^{-1}, reactions (a) and (b) seem to be the most relevant and these call for

$$\nu_s \pm \nu_w = \nu_r$$

If this is correct it means that some of the maxima (and minima) in Fig. 1 and 2 coincide directly with receptor band frequencies while others are satellites which owe their osmic activity to the intervention of water molecules. The separation of a satellite from its primary should correspond to one of the more intense lines in the absorption spectrum of water vapor.

To test this we used a frequency scale on which the positions of the intense water lines were marked and applied it in turn in the positive and negative directions to the peak density plot, sliding it along so as to maximize the number of coincidences (within ± 3 cm^{-1}) between the water lines and the maxima (or minima) in the plot. The result suggests, although inconclusively, that the many maxima and minima in the peak density plots may arise from the participation of water molecules in the stimulus-receptor interaction. If further study shows that water vapor is indeed an energy-transfer catalyst in olfactory stimulation it will call for a new interpretation of the appearance of positive and negative bands in the framework of the vibrational hypothesis.

FREQUENCY BAND WIDTH OF THE RECEPTOR ELEMENT

While examination of the spectra of olfactorily stimulating substances may reveal something about the effective band width of the stimulus, it does not follow that the band width of the receptor is necessarily the same. Absorption spectrophotometry is at best an imperfect tool if only because the dynamic range of the (linear) apparatus is of the order of 10 while that of the (logarithmic) nose is above 10^6 (Fieldner et al. 1931). Furthermore, the need to make the measurements while the substances are dissolved in such non-polar solvents as benzene may perturb the far infrared frequencies to an unknown and variable extent, though the indications are that the peak displacements will seldom exceed 5 cm^{-1};

The spectroscopically observed infrared peaks in organic compounds are often wider than 30 cm^{-1}. Nevertheless, in the peak density plots shown in Fig. 1 and 2 the maxima and minima are sharp enough to suggest that it is the receptor bands which determine the band width for the process of stimulus - receptor interaction. In particular, using a moving 7 cm^{-1} wide interval in preparing the peak density plots gave nulls at certain frequencies for all the responses so far studied. For a wider moving interval of 10 cm^{-1} these nulls largely disappear. It may be concluded therefore that the receptor band widths are close to 7 cm^{-1} for both positive and negative bands, more especially as the distinction between them may be made at the synaptic level rather than at the level of the primary interaction with the stimulus.

This relatively small band width is consistent with the closeness to resonance required for efficient vibrational energy exchange between infrared-active vibrations. The root-mean-square velocity of a molecule with a molecular weight of 200, in equilibrium at 310 K is 2×10^4 cm/sec which can provide effective collisions for energy mismatches up to about 10 cm^{-1} (cf. Yardley. 1969, who, although dealing with a specific case, indicates the essential relations between the molecular parameters and the transfer probability).

EFFECT OF ISOTOPIC SUBSTITUTION

It is often suggested that if the vibrational hypothesis is valid, isotopic substitution ought to affect the odors of compounds. Qualitatively, the argument is this : inasmuch as the frequency of the harmonic oscillator depends upon the restoring force and the masses of the moving parts, substituting a heavier or lighter isotope should have little effect on the forces but should affect the frequency and hence the odor because of the change in mass. Quantitatively for a simple harmonic oscillator the frequency ν is given by

$$\nu = \frac{1}{2\pi} . (\frac{k}{\mu})^{1/2}$$

where k is the restoring force constant and μ is the "reduced mass" of the system, that is,

$$\mu = \frac{m_1 m_2}{m_1 + m_2}$$

where m_1 and m_2 are the masses of the relatively moving parts. Since the expression for the frequency involves $\mu^{1/2}$, it is clear that relatively large changes in m_1 or m_2 will be needed to have in important effect on the frequency and therefore replacing ^{12}C by ^{14}C in a group such as -CH$_3$ will make a difference comparable to deuteration and give a frequency change of at most 5 per cent.

Young et al. (1948) compared the frequencies of the -OH and -OD stretching modes of normal and deuterated butyl alcohol both of which are well above 2000 cm^{-1} and therefore entirely irrelevant to olfaction on quantum grounds.

More recently, Doolittle et al. (1968) deuterated a well-known melon fly attractant 4-(p-hydroxyphenyl)-2-butanone acetate at various locations and reported no change in the insect-attracting properties. The effect of the deuteration on the lowest frequencies was relatively small, amounting to no more than 6 or 7 cm^{-1} in peaks near 225 and 199 cm^{-1}, or about 3 per cent, so that it is unlikely that the frequency was necessarily shifted out of the relevant osmically active band.

In one case isotopic substitution appears likely to alter the vibrational frequency enough to affect the odor. Ordinary naphthalene has far infrared frequencies at 363 and 183 cm^{-1} which are shifted to 331 and 169 cm^{-1} in the fully deuterated compound (Chantry et al. 1964). A triple-blind comparison of the odors using six observers showed that four were able to distinguish the deuterated compound from the normal naphthalene (Stuart, 1965).

ODOR AND OPTICAL ACTIVITY

The question of whether optical antipodes can be discriminated olfactorily has been the subject of much discussion. In a critical review of the data, Naves (1957) concluded that the most that had been demonstrated with adequately pure samples was a slight difference in note and strength. Also, in the few cases where the dextro- and levo-forms of an insect attractant have been tested the response was the same to both forms (Jacobson, 1962; Schneider, 1963). What is certain is that no case is known where one antipode has an odor and the other has none.

Since the vibrational frequencies and dipole moments of mirror-image isomers are necessarily the same, the possibility of odor differences requires some comment.

It is well known that biochemical processes and systems involve interactions between dissymmetric molecules whose chirality or "handedness" is often of major importance. It is therefore a matter of some surprise that it has been so difficult to demonstrate such an effect in olfactory perception when very marked relations have been found between chirality and taste. (Bentley, 1969). The explanation probably lies in the nature of the vibrational interaction which, because it takes place between molecules which are separated by a distance of a few Å is less subject to the effects of chirality than it would be if the approach was close enough to be regarded as chemisorption.

HUMAN OLFACTORY THRESHOLDS

If the best olfactory threshold given by Laffort (1963) is plotted against the lowest fundamental infrared-active frequency, the points fall on or very near a straight line.

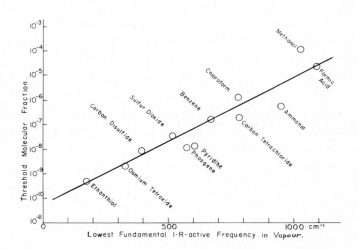

Fig. 3. Plot of the threshold molecular fraction against the lowest fundamental vibrational frequency infrared-active in the vapor state.

Since the figure includes such disparate substances as osmium tetroxide, heptane and formic acid, it seems clear that the threshold is related to a single parameter of the molecule – its vibrational behavior – and is independent of the chemical class to which the molecule belongs.

CONCLUSION

The examination of infrared absorption spectra of compounds which evoke specific olfactory responses has confirmed an association between the two. A particular stimulus pattern may depend on the absence of certain frequencies as much as on the presence of others.

Whereas the vibrational hypothesis was originally intended to apply to the problem of odor quality and discrimination, it is now evident that the same hypothesis provides a direct correlation with the olfactory threshold.

What other model of human olfaction can provide an equally quantitative insight into the molecular mechanisms of olfactory discrimination and sensitivity?

ACKNOWLEDGMENT

The authors acknowledge with gratitude a grant from the Donner Canadian Foundation.

REFERENCES

Bentley, R. 1969. "Molecular Asymmetry in Biology", Vol. I. Academic Press.

Beroza, M. and Green, N. 1963. "Materials Tested as Insect Attractants". U.S.D.A., Agriculture Handbook No. 239.

Boeckh, J. 1965. Inhibition and Excitation of Single Insect Olfactory Receptors and their Role as a Primary Sensory Code. Proc. 2nd. Int. Symp. Olf. & Taste, Tokyo. Pergamon, 1967.

Chantry, G.W., Anderson, A. and Gebbie, H.A. 1964. Far Infrared Spectra of Naphthalene-H_8 and -D_8. Spectrochim. Acta, 20, (9), 1465-1466.

Cottrell, T.L. and McCoubrey, J.C. 1961. "Molecular Energy Transfer in Gases". Chap. 6, Butterworth, London.

Doolittle, R.E., Beroza, M., Keiser, I. and Schneider, E.L. 1968. Deuteration of the Melon Fly Attractant, 'Cue-Lure', and its Effect on Olfactory Response and Infrared Absorption". J. Insect Physiol. 14, 1697-1712.

Fieldner, A.C., Sayers, R.R., Yant, W.P., Katz, S.H., Shohan, J.B. and Leitch, R.D. 1931. Warning Agents for Fuel Gases. U.S. Bureau of Mines, Monograph No. 4.

Gesteland, R.C., Lettvin, J.Y., Pitts, W.H. and Rojas, A. 1962. Odor Specificities of the Frog's Olfactory Receptors. Proc. 1st Int. Symp. Olf. & Taste, Stockholm, 19-34. MacMillan, 1963.

Gesteland, R.C., Lettvin, J.Y. and Pitts, W.H. 1965. Chemical Transmission in the Nose of the Frog. J. Physiol. (London), 181 (1), 525-559.

Hainer, R.M., Emslie, A.G. and Jacobson, A. 1954. An Information Theory of Olfaction. Ann. N.Y. Acad. Sci. 58, Art. 2, 158-174.

Jacobson. M. 1962. Insect Sex Attractants. III. Optical Resolution of dl-10-acetoxy-cis-7-hexadecen-1-ol. J. Org. Chem. 27, 2670-2671.

Laffort, P. 1963. Essai de Standardisation des Seuils Olfactifs Humains pour 192 Corps Purs. Arch. Sci. Physiol. 17, (1), 75-105.

Naves, Y.R. 1957. The Relationship Between the Stereochemistry and Odorous Properties of Organic Substances. In "Molecular Structure and Organoleptic Quality". S.C.I. Monograph No. 1, pp 38-51.

Schneider, D. 1962. Electrophysiological Investigation of Insect Olfaction. Proc. 1st Int. Symp. Olf. & Taste, Stockholm, 85-103. MacMillan, 1963.

Stevens, B. 1967. "Collisional Activation in Gases". Chap. 2, Pergamon.

Sturart, R.S. 1965. Private communication. See A. Demerdache, Proc. 2nd Int. Symp. Olf. & Taste, Tokyo, 1965, p. 135. Pergamon, 1967.

Wright, R.H. 1954. Odor and Molecular Vibration. I. Quantum and Thermodynamic Considerations. *J. Appl. Chem. 4,* 611-615.

Wright, R.H. 1966. Primary Odors and Insect Attraction. *Can. Ent. 98,* (10), 1083-1093.

Wright, R.H. and Robson, A. 1969. Basis of Odor Specificity : Homologs of Benzaldehyde and Nitrobenzene. *Nature 222,* 290-292.

Wright, R.H. and Burgess, R.E. 1969. Musk Odor and Far Infrared Vibration Frequencies. *Nature 224,* 1033-1035.

Wright, R.H. 1970. Insect Attraction – Wasps. *Israel Jl. Entomol.*

Wright, R.H. and Burgess, R.E. 1969. Specific Physicochemical Mechanisms of Olfactory Stimulation. Proc. Ciba Found. Symp. J. & A. Churchill, 1970.

Yardley, J.T. 1969. Non-Resonant Vibration-to Vibration Energy Transfer due to Dipole-Dipole Interaction. *J. Chem. Phys. 50,* 2464-2466.

Young, C.W., Pletcher, D.E. and Wright, N. 1948. On Olfaction and Infrared Radiation Theories. *Science 108,* 411.

DISCUSSION

Leffingwell : Pure synthetic l-carvone has a spearmint odor whereas pure synthetic d-carvone has a caraway odor (Langenau, 1967). An even more striking example is synthetic l-menthol and d-menthol; these are different both in odor and in threshold level. These facts are in contradiction to your theory.

Wright : Doll and Bournot (1949) purified the d- and l-forms of menthol very carefully and found that they had the same minimum stimulus and the same odor up to twenty times the threshold concentration, after which they diverged. But menthol is not a simple odorant, for it stimulates the cold receptors also, and according to Hensel and Zotterman (1951) this probably involves an enzymatic mechanism and would therefore be different for the two forms of menthol.

In the case of d- and l-carvone, reported by Langenau (1967), he precedes his comment with the remark, "In the case of many aromatic chemicals, optical activity has little effect on the odor", and with this I think everyone must agree. As pointed out in my presentation it is a surprising and therefore a significant fact that there are so few authenticated cases where dextro and levo isomers have appreciably different odors and no case where one is odorous and the other inodorous. To my mind the best explanation is that the process of interaction does not involve a sufficiently close approach of the stimulus molecule to the receptor surface to make the effects of chirality important except in a few exceptional cases.

Beets :   Theimer and McDaniel found that d- and l-α-pinene are distinctly different in odor.

Schneider :   You and many others are working on a system much too complex to make any kind of predictions of the type we heard about. We know from our own and other experiments with humans that we are probably dealing with single molecule-acceptor interactions. What you measure are mass effects. To make predictions from mass effects in gas, liquid, or interfaces, is a great risk. What about molecules which exhibit your specific vibrational properties but do not smell at all?

Dravnieks :   It is extremely difficult to accept that part of the vibrational theory which attempts to explain odor sensing in terms of some unspecified excited pigment system, which, presumably, probes for the presence or absence of olfactorily active frequencies. It is very easy to accept that there exist correlations between odor and frequencies, since frequencies describe molecular structure and, indirectly, all other properties of odorants. Thus, the question is : Do some selected simple properties relate to odors in a simple way? I have mentioned correlations related to odor similarities (see discussion to Moulton's paper, this symposium). The present comment deals with sensitivities. These are best measured in terms of Stevens slopes which represent log (response) vs log (concentration) plots; odor "threshold" is merely an incident where the intensity reaches a level at which, within the context of motivation, chemical and physiological noise during the test and the statistical criterion selected, doubt arises if an odorant is present. Recent work by Laffort (Paris) and us enabled testing whether odor intensity slopes from the literature and our own data can be correlated with simple physicochemical properties of odorants. Regardless of the psychophysical method (category scaling, magnitude estimate, and match vs a reference odorant) used, strong correlations ($p < 0.01$; sometimes $p \ll 0.001$) were found between the slope and : (1) simple polynomial combinations of molvolume, hydrogen bonding, and electronic polarizability parameters, (cf. yesterday's remark to Dr.Moulton's paper) (2) the same parameters combined by Laffort's equation (Laffort, 1969); (3) simple polynomial combinations of gas chromatographic characteristics such as Kovats Indexes and Rohrschneider constants. Significance of the parameters in Laffort's equations is conceptually based on interactions between odorant molecule, mucus water, and proteinic receptors. Thus, although correlating these properties to spectral characteristics would be interesting, simpler properties seem to suffice in explaining many differences in odor sensitivity, at least in humans.

REFERENCES

Doll, W. and Bournot, K. 1949. Odor of optical antipodes. *Pharmazie 4*, 224-227.

Hensel, H. and Zottermann, Y. 1951. Effect of menthol on the thermoreceptors. *Acta Physiol. Scand. 24*, 27-34.

Laffort, P. 1969. "Olfaction and Taste" vol. III, p. 150, ed. Pfaffman, C. Rockefeller University Press, New York.

Langenau, E.E. 1967. Correlation of objective-subjective methods as applied in the perfumery and cosmetics industry. *Amer. Soc. Test. Mater. Spec. Tech. Publ. 440,* 76-86.

PROGRESS TOWARDS SOME DIRECT QUANTITATIVE COMPARISONS OF THE STEREOCHEMICAL AND VIBRATIONAL THEORIES OF ODOR

John E. Amoore

Western Regional Research Laboratory, Agricultural Research Service,
U.S. Department of Agriculture, Albany, California, 94710, U.S.A.

INTRODUCTION

Among the many theories of odor that have been propounded over the years, there are two which because of their longevity, if for no other reason, especially invite comparison. My own stereochemical theory was proposed in 1952, has been expanded in many papers, and was recently summarized in a book (Amoore, 1952, 1970). Dr. R.H. Wright's vibrational theory dates back to 1954, is also the subject of a continuing series of publications, and was described in his book a few years ago (Wright, 1954, 1964). In essence, my theory claims that the size and shape of the molecule govern the type of odor, whereas Wright's theory contends that the inherent vibrational frequencies of the molecule are the causative factor. For each theory a substantial amount of experimental support has been adduced by its author or by other workers. Often this support is quantitative in nature, and impressive degrees of statistical significance have been achieved (Amoore and Venstrom, 1967; Wright et al., 1964). Nevertheless, up to the present, there has been no direct quantitative comparison of the two theories, when facing the same set of olfactory data. This is the situation which Dr. Wright and myself, in a lengthy series of personal communications, have resolved to rectify.

It must be emphasized at the outset that no single paper, such as the present, could be expected to reach a definitive agreed conclusion. Indeed, due to various unavoidable delays, the necessary data are not yet completely assembled, let alone digested. Rather this paper should be regarded as the opening statement in a "debate", a debate which may extend over several alternate publications by the principals and which could only be concluded, if at all, by the weight of experimental evidence.

Dr. Wright and I have agreed to consider three basic sets of olfactory data. The first set consists of 15 lower fatty acids, which exhibit in varying degrees a "sweaty" odor quality. The "sweatiness" of the odor was measured by the amount of olfactory detection threshold deficiency observed in a group of ten persons having a specific anosmia towards the sweaty odor. Isovaleric acid was established as the typical primary odorant. The quantitative values (in binary dilution steps) have been published (Amoore et al., 1968).

The second set is composed of 27 homologous and isomeric alkyl-substituted benzaldehydes and nitrobenzenes. To differing extents the "almond" odor character is exhibited. The "almondiness" is determined by collecting judgments of odor similarity to the type compound, benzaldehyde itself, from a panel of observers. The similarity values of 25 judges for 9 of these compounds have been published (Klouwen and Ruys, 1963). Through the generosity of the Naarden Company, where this research was originally carried out, samples of all but one of the 27 compounds have been collected, and they will soon be submitted to odor similarity testing by my panel of 30 observers.

The third set of data concerns the olfactory sense of the ant *Iridomyrmex pruinosus,* which produces 2-heptanone as its pheromone signaling alarm. The activities of 50 related ketones and 35 non-ketones were tested as releasers of the alarm reaction by Prof. M.S. Blum of the University of Georgia. The alarm ratings for the ketones have been published (Blum et al., 1966), and those for the non-ketones are being prepared for publication. Samples of 62 of these 85 compounds have been reassembled for measurements of their far infrared spectra by Dr. Wright.

It should be noted that the first set of compounds (fatty acids) has been closely scrutinized (Amoore, 1967), whereas the second set (benzaldehydes) has been the subject of thorough study by Wright and Robson (1969). Both investigations revealed significant correlations between the olfactory data, and either stereochemical or vibrational assessments respectively. The point at issue is this : Can the relative success of the two theories be subjected to quantitative comparison, and if so, which yields the better correlation?

The experimental data for the third set of compounds (ketones) were observed on a test organism (the ant) which reacts instinctively, and were measured by an entomologist who had no preconceived opinion on the value of these or any other theories of odor. The alarm pheromone data may therefore provide an independent check on the proceedings.

We plan to arrange the confrontation of the theories, as far as possible, according to the provisions of the scientific method. The preliminary phase of collecting experimental olfactory data has been virtually completed (anosmia values, odor similarities and alarm ratings). The formulation of competing hypotheses has also been done; indeed, they have been perhaps prematurely elevated to the status of "theories" (stereochemical theory and vibrational theory). Now comes the testing of the hypotheses, to see how well each can explain the existing olfactory data.

Each theory suggests certain physico-chemical properties of the odorous molecule, properties that are believed to be related to its odor. In the stereochemical theory the size and shape of the molecules are regarded as the principal factors. In the vibrational theory the frequency and intensity of the intra-molecular oscillations are considered to be of paramount importance. For both theories, the requisite chemical properties are open to experimental measurement (see Methods). Approximate molecular sizes and shapes can be obtained from photographs of the scale molecular models (Amoore, 1965). The molecular vibrations themselves are rather inaccessible, but Wright and Robson (1969) use infrared-active vibrational frequencies from far infrared spectra.

Thus each theory can furnish a series of photographs indicating for each compound under consideration the relevant molecular shapes or molecular vibrations. These photographs are not the only, nor perhaps the best, basis for assessing the respective molecular properties. They do, however, appear to provide the material for a straightforward, rather elementary comparison of the two theories. Thus the photographs themselves could be measured and assessed, by manual methods in the first instance. Later a scanning computer might be employed to provide an independent corroboration of the manual measurements. Prof. G. Palmieri and his collaborators at the University of Genoa have kindly agreed to submit the photographs to their PAPA pattern recognition machine (cf. Palmieri and Wanke, 1968).

Both manual and computer procedures may supply numerical assessments of the molecular shapes or vibrations, i.e. the physico-chemical properties. These results may be compared with the olfactory data, i.e. the biological properties of the compounds. For this purpose graphs may be plotted, showing how the chemical and biological data compare for a given set of compounds. Visual inspection should indicate at a glance whether a systematic trend exists in the distribution of the points, and which theory gives the better concordance with the experimental facts. For an assessment of the concordance, the correlation coefficient (**r**) of statistics may be calculated. This would appear to provide as fair a yardstick as any for measuring the overall success of the stereochemical and vibrational theories of odor, in explaining existing experimental data.

However, for a really severe test of a theory, it is necessary to go beyond existing data and enter the realm of prediction and confirmation of previously unknown facts. This is the ultimate step of the scientific method. It is necessary because any hypothesis can be made sufficiently elaborate to "explain" any given collection of known facts, but it may begin to break down when extrapolated into terra incognita. Success at this stage would justify the promotion of a hypothesis to the rank of theory. A bold foray into predicting alarm activities in 15 untested compounds has been made by Dr. Wright through application of his vibrational theory. With these predictions my stereochemical theory is in substantial disagreement. Time will tell which, if either, of us is correct!

METHODS

Olfactory data

The procedures for making the quantitative olfactory measurements have been described elsewhere. The 15 fatty acids with the sweaty odor were graded by the specific anosmia method (Amoore et al., 1968). Nine of the 27 benzaldehydes and nitrobenzenes (all the *ortho*-substituted isomers) were assessed by a similarity method. The judges were asked to decide whether the test compound more resembled its parent hydrocarbon or benzaldehyde (or nitrobenzene) itself (Klouwen and Ruys, 1963). Later I intend to measure the odor similarities of the whole series to benzaldehyde alone, by using another similarity method (Amoore and Venstrom, 1966). The methods for rating the alarm-releasing activity of various compound on *Iridomyrmex pruinosus* have been explained before (Blum et al., 1966, Amoore et al., 1969).

All the olfactory data are recorded on numerical scales. The specific anosmia values are in the range 0 to 7 binary dilution steps of threshold deficiency, with an average standard error of the mean of ± 0.65 step (Amoore et al., 1968). The results of Klouwen and Ruys (1963) are based on evaluations of 25 judges who voted on whether the almond character dominated; no estimate of variability is available. My panel's estimates of odor similarity are on a scale from 0 to 8, in which the standard error of the mean is about ± 0.35 step (Amoore and Venstrom, 1966). The ant alarm activities were reported on an integral rating scale from 0 to 5 steps, where the standard error of the mean is substantially less than ± 0.5 step (Amoore et al., 1969).

Although the olfactory measurements include appreciable experimental error, the selfsame tables of olfactory data are to be used for comparison with both theories. It is to be expected that error in the olfactory measurements will lower the final correlation coefficients obtained, but there is no reason to think that the error could favor one theory over the other. Hence these olfactory data, error notwithstanding, constitute a fair test of the theories.

Stereochemical data

The procedures for producing the molecular silhouette photographs have been described (Amoore, 1964). Three orthogonal views of each molecule are obtained, and all three are used when comparing molecular shapes. A fixed set of rules is employed to decide on the most likely conformation of the molecule, and to select the standard orientation of axes for photography. Representative molecular model silhouettes are shown in Fig. 1.

Fig. 1. Molecular silhouette photographs of 15 fatty acids. These pictures constitute the raw material for the stereochemical assessment of the compounds.

The molecular silhouettes may be compared (Amoore, 1965) by means of radial measurements from the center of gravity of each silhouette to its periphery (Fig. 2). The algebraic differences in length (overlap) of the corresponding radii of two molecules are obtained and summed over

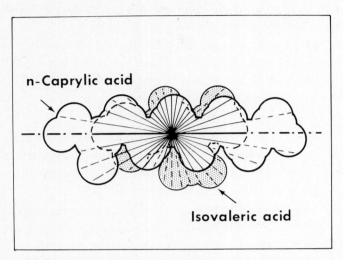

Fig. 2. Comparing the silhouettes of molecular models by the manual method, for the stereochemical theory. Dashed lines indicate the differences (overlap) of corresponding radii.

all three molecular aspects. In the present study, the average difference ($\overline{\Delta}$) is subtracted from a convenient larger constant (2.0 Å). This yields a measure of the **similarity** between the molecular silhouettes. (The reciprocal form previously employed (Amoore et al., 1967) accentuated an undesirable distortion of the curve at high molecular similarities).

The above manual method is slow but accurate. It has been superseded by a much faster procedure which employs a scanning computer (Amoore et al., 1967). This instrument calculates the similarities between two pictures by scanning them with a reproducible set of random lines, then comparing the intersections made by each line with salient features in the two fields of view. The PAPA machine avoids the risk of arithmetical errors. Furthermore, being operated by independent investigators who do not know either the olfactory data or the manual results, the PAPA machine brings impartiality to the comparison of theories.

Vibrational data

Wright and Robson (1969) measure the far infrared absorption spectrum of each compound with a Perkin-Elmer model 301 far infrared grating spectrophotometer. The samples required are usually of the order of $\frac{1}{4}$ ml. The samples were dissolved in n-heptane at 1 to 10 % concentration. The frequency range recorded is from 500 to 80 cm^{-1} (reciprocal centimeters or wave numbers). Wright (1954 a, 1954 b, 1964 b) has explained earlier the theoretical reasons for considering that this frequency range is the most relevant for reflecting those molecular properties that are linked to odor qualities.

The spectra usually contain a number of distinct peaks, whose frequencies can be accurately determined by visually inspecting the curve in conjunction with the instrumental scale on the chart paper. The positions of these absorption maxima constitute the raw material for Wright's deductions according to the vibrational theory. Sometimes a definite shoulder is visible on the side of a peak, and its central frequency can be estimated with fair confidence. Representative spectra have been published (Wright and Robson, 1969).

It should be noted that Wright generally does not employ the relative absorption intensity information contained in the spectrophotometric record. In fact he has stated that "Peak heights vary and are of uncertain relevance, and therefore frequencies only are plotted ...." (Wright and Robson, 1969). For his calculations, the peak frequencies alone are sufficient indicators of whether a given compound will fall into one odor category or another.

Nevertheless, the present comparison of theories is directed beyond the qualitative stage towards a quantitative goal. The questioner will not rest content with a yes or no answer to the query " Is the molecule expected to have an almond odor? " We need a quantitative estimate of "What degree of almondiness should the molecule exhibit? "

I myself had hoped that it might be possible to use the absorption peak heights, as well as their positions, in an attempt to develop comparable quantitative data for assessing the two theories. The spectral curves superficially appeared to me to invite comparison by, for example, the measurement of corresponding ordinates. Perhaps even the area below the curve could be blacked in, and the resulting silhouettes presented to the PAPA scanning computer.

Unfortunately it transpired that the published spectra do not lend themselves to this elementary treatment. The spectrophotometer was operated by Dr. Wright in such a way as to obtain the clearest resolution of the absorption peaks, with accurate identification of the peak frequencies. To this end, the molar concentration of the odorant in the heptane, the sensitivity of the instrument, and the height of the curve above the base line, were variously adjusted to yield the most informative spectrum. Informative, that is, with respect to frequency; but in maximizing the frequency information, the intensity information is minimized, and certainly cannot be compared from one compound to another, not in the spectra presently available.

Hence I will await the development by Dr. Wright of a quantitative method for comparing the spectra of different compounds, based exclusively on their frequencies, and ignoring the intensities of the absorption bands. The method of choice should ideally yield a numerical assessment of the similarity of each spectrum to that of the reference compound, such as isovaleric acid or benzaldehyde. Then a parallel test of our two theories would become possible, by correlating the results of each with the same set of olfactory data.

RESULTS AND DISCUSSION

Sweaty odors of fatty acids : stereochemical theory

I began the comparison between the two theories with the 15 carboxylic acids. The compounds studied were the normal saturated fatty acids from one through ten carbon atoms, plus isobutyric acid, three more valeric acid isomers, and isocaproic acid. The physico-chemical data offered by the stereochemical theory are displayed pictorially in Fig. 1. The stereochemical theory starts from the molecular silhouette photographs, the top-view silhouettes only being shown. The scale is in Å units.

In Fig. 1 the photograph for isovaleric acid is singled out by a frame, because the olfactory experiments indicate that this compound most closely epitomizes the purest primary odor sensation, and most nearly possesses the ideal molecular properties of eliciting the sweaty odor (Amoore, 1967). Hence the physical properties of isovaleric acid should be employed as a standard with which the properties of the other acids are compared.

That the five acids in the last column of Fig. 1 closely resemble each other in molecular shape is obvious to the eye, and it should come as no surprise to learn that they all have highly sweaty odors. As a matter of fact, several of the straight-chain acids are also very sweaty, particularly those with 4 to 7 carbon atoms (n-butyric through n-enanthic) because they approach isovaleric acid in molecular size.

Fig. 3 shows the results obtained by the stereochemical theory on the 15 fatty acids. The

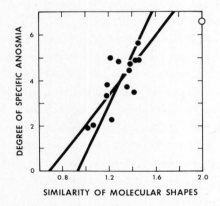

Fig. 3. The stereochemical theory provides a satisfactory correlation with the observed odor characteristics of 14 fatty acids. (The larger open circle represents the primary odorant, isovaleric acid itself).

y-axis represents the olfactory data that we are seeking to explain, the degrees of specific anosmia exhibited by each acid (Amoore et al., 1968). The scale is in binary concentration steps difference in odor detection threshold between normal observers and specific anosmics. These are the experimental olfactory values that must be matched by means of physicochemical measurements on the same molecules, disposed along the x-axis.

As can be seen in Fig. 3 the manual measurements of molecular similarity correlate quite well with the anosmia values. The trend of the experimental points is definitely from the lower left towards the upper right of the graph. Both regression lines are drawn, assuming first that molecular shape, and then that specific anosmia, is the independent variable. This was done because there is no justification for assuming that all the experimental error is associated with one variable exclusively. The pair of lines also serves the useful purpose of indicating at a glance the degree of correlation. The closer the lines converge, the better the correlation coefficient, to a maximum of $r = 1.0$ for a straight line graph. The more divergent the lines, the worse the correlation, to a minimum of $r = 0.0$ for complete randomness.

The regression lines in Fig. 3 converge fairly satisfactorily. The correlation coefficient achieved by the stereochemical theory is $r = 0.77$, highly significant at a probability level of $P \sim 0.001$ for $N = 14$ acids. (In all the statistical calculations in this paper, the standard compound, in this case isovaleric acid, is excluded on the grounds that it is inevitably identical with itself.) Hence, the more closely the molecular silhouette of a fatty acid resembles isovaleric acid in size and shape, the greater will be the degree of specific anosmia (to sweatiness) that it reveals. (This conclusion, with substantially the same correlation coefficient, has been reported and illustrated in earlier publications, with a somewhat different arithmetical function on the x-axis, e.g. Amoore, 1967).

Sweaty odors of fatty acids : vibrational theory

Now it is the turn of the vibrational theory to develop a correlation with the olfactory data. Wright has measured the far infrared spectra of all the 15 fatty acids, and he will be publishing them in due course. What we need is some measure of molecular vibrational similarity that can be placed along the x-axis of e.g. Fig. 3, so that a comparable graph can be plotted. Then from the distribution of the experimental points, and from the resulting correlation coefficient, we will be able to judge which theory can give the better explanation for the observed degrees of specific anosmia.

A complication that Dr. Wright has encountered is that not all regions of the far infrared spectrum are relevant for a particular odor. I quote once more from his paper (Wright and Robson, 1969) : "Moreover, not all the molecular vibrational movements are osmically active ...." This qualification unfortunately means that it is not possible to make deductions from the spectrum of a single known good representative of a class of odorants, such as isovaleric acid, because there is no a priori means of telling which of its six obvious peaks and two prominent shoulders are osmically active. It now becomes necessary to examine the spectra of several compounds known to possess the target odor in substantial degree, so as to establish which peak or group of peaks may be common to all compounds with that odor.

Apparently in this series of 15 fatty acids, a single feature runs through the spectra of all the sweatier-smelling members of the group — "the presence of a strong absorption in the vicinity of 360 cm^{-1}, plus or minus 15 or 20 cm^{-1}" (letter from R.H. Wright, 31 December 1969).

From my own inspection of the spectra Wright kindly sent to me, this remark seems intuitively true. However, I feel that it ought to be accompanied by a reminder that the strength (i.e. intensity) of the absorption should not be considered relevant in comparing the available spectra of different compounds, because the spectrophotometric procedure was not standardized (see Methods). To my unpractised eye, a recognizable peak or shoulder does appear within the specified frequency range in 13 of the 15 acids. The presence of an absorption peak with this frequency may be a qualitative indicator of a potentially sweaty odor, but we still need some way of abstracting from the spectra a quantitative measure of the degree of sweatiness to be expected. I look to Dr. Wright to work out the appropriate calculations. Until then, our two theories cannot be quantitatively compared.

Independent tests by the PAPA machine on the fatty acids

To obtain a "second opinion" on the stereochemical data, the photographs of the molecular models for the fatty acids were also submitted for examination by the PAPA scanning computer. As previously published (Amoore et al., 1967), the molecular shape similarities correlate well with the specific anosmia data, $r = 0.80$, highly significant at $P < 0.001$. This value was obtained when the PAPA machine used the center of gravity of the molecular silhouette as the datum point. Substantially the same result was achieved if instead the position of the functional group (carboxyl) was used as datum point in placing the silhouette photograph in the camera field of view. Hence the use of the PAPA machine supports the conclusion reached from manual measurements, that the closer a carboxylic acid resembles isovaleric acid in molecular shape, the more sweat-like its odor will be.

Once the machine had been optimally adjusted on the 15 fatty acids, those same machine settings were recorded, and are used in all subsequent tests of molecular silhouettes, whether they be ether, camphor, musky, floral or minty odors (Amoore et al., 1967), the almond odor, the ant alarm pheromone activities (Amoore et al., 1969) or any other problem of chemical constitution and biological activity that comes along. Only the single standard of comparison is changed for each class of odor (isovaleric acid molecular silhouette for sweaty, 1,8-cineole for camphoraceous, 2-heptanone for *Iridomyrmex pruinosus,* etc.).

The PAPA machine is unfortunately not suitable for comparing the infrared spectra directly , because the scanner is unavoidably influenced by the heights of the peaks (i.e. their intensities) as well as their positions (frequencies). Bearing in mind that the intensity information is inadmissable on these spectra, the machine in this configuration cannot yield informative results. Perhaps the spectra should be re-drawn with only the positions of the peaks indicated, e.g. by uniform vertical lines. Alternatively the PAPA machine, which is capable of considerable learning capacity, might be "taught" to disregard intensity data and concentrate only on the position of the peaks in the spectra.

Almond odors of benzaldehydes (incomplete data)

Quantitative data on the almond character in this series are available for only 10 compounds (Klouwen and Ruys, 1963). These are benzaldehyde itself, nitrobenzene, and their *ortho*-substituted monoalkyl derivatives, with methyl, ethyl, isopropyl or tertiary butyl in the *ortho* position. The odor similarity values recorded in Table 1 were obtained from the votes of a panel of 25 observers (see Methods).

Using benzaldehyde itself as the standard of molecular shape, and the same PAPA machine settings as for the fatty acids, the molecular shapes of the other nine compounds were assessed. The values obtained are recorded in Table 1 and clearly run parallel with the odor values. The correlation coefficients were $r = 0.89$ by the center of gravity method and $r = 0.92$ by the functional group method, both highly significant at $P \sim 0.001$ with $N = 9$ compounds.

Evidently the stereochemical theory has no problem accommodating the data from the unfamiliar field of the almond odors. In fact the correlation coefficient is so high that the vibrational theory will have to develop some impressive output to compete. For the almond odor, Wright postulates a much more complex set of criteria than the single band that sufficed for the sweaty odor. Apparently the almond character is associated with peaks in the 175, 225 and 345 cm^{-1} bands, and perhaps a scarcity of peaks in the 265 and 310 cm^{-1} bands (Wright and Robson, 1969).

The infrared spectra for this series have been published (Wright and Robson, 1969) and an analysis of the distribution of spectral frequencies among these bands has been made. The results permit a very significant distinction between the group of 16 compounds having an almond odor, and a group of 31 control compounds that have other odors. However, the distinction is essentially

Table 1. Correlation between molecular shape and almond odor character.

Compound	Molecular shape*	Odor similarity **
Benzaldehyde (standard)	(1206)	(25)
o-Methylbenzaldehyde	1090	25
o-Ethylbenzaldehyde	1031	16.5
o-Isopropylbenzaldehyde	1027	3
o-t-Butylbenzaldehyde	1008	0.5
Nitrobenzene	1125	25
o-Methylnitrobenzene	1081	24
o-Ethylnitrobenzene	1033	14.5
o-Isopropylnitrobenzene	1015	4
o-t-Butylnitrobenzene	988	4

* Similarity values to benzaldehyde, using PAPA machine, center of
gravity method.

** Resemblance to benzaldehyde (or nitrobenzene), votes of 25 observers.

qualitative, and between groups. It does not at present lead to any quantitative estimate of the relative degree of almond character to be expected from each individual compound. Hence no direct comparison of the theories is yet possible.

The complete set of stereochemical assessments on the 27 compounds has been made by the PAPA machine and submitted to Dr. Wright. (The other 17 compounds are the corresponding *meta* and *para* isomers, plus benzonitrile). Further consideration of the almond series must await the collection of a complete set of quantitative olfactory data for comparison with the two theories.

Ant alarm pheromone activity of ketones

Blum et al. (1966) of the University of Georgia have obtained numerical ratings on the effectiveness of 50 ketones and 35 non-ketones as releasers of the alarm reaction in *Iridomyrmex pruinosus*. The natural pheromone of this species, 2-heptanone, was used as the standard for comparisons of molecular shapes. As previously reported, the measurements by the PAPA computer correlated quite well with the alarm ratings (Amoore et al., 1969). With the functional group as the datum point and the same PAPA machine settings as for the fatty acids, the 49 ketones gave $r = 0.57$ and the 35 non-ketones yielded $r = 0.81$ (both values extremely significant, with P about 10^{-5} and 10^{-10} respectively). Use of the center of gravity of the molecular silhouettes gave comparable results. Evidently the stereochemical theory, developed for the human sense of smell, is adaptable to explaining insect pheromone specificity.

In recent months 62 of the original 85 test compounds have been re-assembled, and Dr. Wright has run their far infrared absorption spectra (un-published work). He has compared the spectra of various alarm-active compounds and discerned certain qualitative characteristics in the distribution of absorption maxima, which distinguish them from inactive compounds.

Predictions of alarm activity among untested compounds

To explain a given batch of existing experimental data in any scientific discipline is not too difficult if one has freedom to formulate a sufficiently complex hypothesis. In fact, several possible solutions may spring to mind, and they could be equally successful in correlating with the known data. A real distinction between the merits of rival hypotheses may not be discerned until one attempts to apply them to making predictions of unknown facts.

Dr. Wright has taken the first courageous step in this direction. As mentioned above, he has worked out certain properties of the infrared absorption spectra that appear to characterize compounds with high activity as releasers of alarm in *Iridomyrmex pruinosus*. He has not had the time to publish the details of how he identifies the relevant bands for known active compounds, or how their presence should be assessed in new, untested compounds. Nevertheless, that does not affect the following contest. Suffice it to say that he himself searched through his files of more than 500 reference spectra, applying his own criteria, looking for substances that have the same general spectral character as the known alarm-active compounds. He came up with a list of 15 untested substances (Table 2) that he expects to act as alarm releasers for *Iridomyrmex pruinosus* (letter from Dr. R.H. Wright to Prof. M.S. Blum, 16 February 1970). To quote from that letter: "I believe that the test of any theory is its ability to make verifiable predictions. Therefore while I would not expect all the above to act as alarm pheromones for *Iridomyrmex*, I will be disappointed if a substantial proportion (say one half) do not."

Table 2 therefore constitutes a list of predictions made by means of the vibrational theory, and by its author himself. I have decided to submit the same 15 compounds to an examination by means of my stereochemical theory. Accordingly their molecular silhouettes were photographed and sent to the PAPA scanning computer for comparison with the silhouette of the natural pheromone, 2-heptanone. As datum point, the functional group of each compound was used or its most polar group if more than one is present. (Because trimethylbenzene has no functional group, the center of gravity was used instead.) The degree of similarity in molecular shape between each compound and 2-heptanone, as printed out by the PAPA computer, is entered in Table 2.

To predict the expected alarm activity rating for each compound, it is necessary to have a calibration curve relating the measured PAPA molecular similarities to the observed ant alarm activities for a fair number of compounds. All but one of Wright's predicted substances are non-ketones. Data obtained on a collection of 35 non-ketones in the earlier study (Amoore et al., 1969) have been plotted out in Fig. 4. For a definite calibration curve, I adopted the dashed line bisecting the angle between the regression lines (because neither variable is sufficiently free from experimental error to be considered independent). By using this line, the PAPA molecular similarity data on Wright's 15 compounds were converted to predicted alarm activities, which are listed in the last column of Table 2.

Table 2. Compounds predicted by the vibrational theory to show ant alarm activity

Compound	Molecular shape *	Sterically predicted rating **
Butyl benzoate	106.5	1.9
Triethylamine	102.8	1.3
a-Ionone	103.8	1.5
Safrole	109.9	2.4
1,2,4-Trimethylbenzene	103.0	1.4
Eugenol	104.3	1.5
2,6-Dimethyl-5-heptenal	100.5	1.0
6-Methyl-3-cyclohexene-1-carboxylic acid, 1-methylpropyl ester	104.0	1.5
6-Methyl-3-cyclohexene-1-carboxylic acid, butyl ester	105.2	1.7
3,7,11-Trimethyl-1,6,10-dodecatrien-3-yl acetate	102.3	1.2
6-Methylheptyl acetate	109.1	2.3
Butyl propionate	112.6	2.8
cis-9-Tetradecenyl acetate	96.4	0.4
1-Methyl-2-butenyl 2-methylbutyrate	104.7	1.6
4-Methylpent-2-yl a-chloropropionate	107.5	2.0

*Similarity values to 2-heptanone, using PAPA machine, functional group method.
**Predicted by stereochemical theory, using calibration curve of Fig. 4.

It is apparent that among Wright's compounds, only n-butyl propionate (predicted rating nearly 3) is a serious contender for the alarm activity of 2-heptanone, according to the stereochemical theory. A few others may achieve slight activity, but at least 8 as indicated should register only rating 1·5 or less, which represents virtual or complete inactivity on Blum's rating scale.

Prof. Blum is hoping to run the bio-assays with these 15 compounds on *Iridomyrmex pruinosus* during the 1970 summer season when the ants are active. He has not been informed of the predictions made by me in this paper, so his data, when they become available, should permit an independent test of the two theories.

Considering that both the vibrational and stereochemical theories have "gone out on a limb" in making predictions which differ rather radically from each other, we are awaiting Prof. Blum's results with more than passing interest, as a means of judging the relative merits of the two theories.

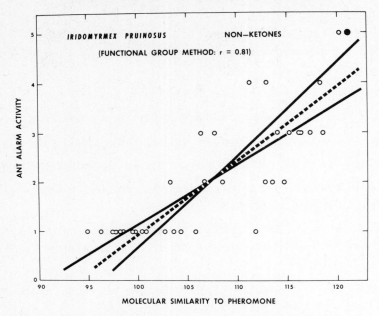

Fig. 4. Existing data on the alarm pheromone activity ratings of 35
non-ketones correlated quite well with the molecular silhouette
measurements. The dashed line was used for predicting the expected activities of untested compounds, according to the stereochemical theory (see Table 2). The larger black circle represents
2-heptanone itself, the natural pheromone.

CONCLUSION

This is a progress report, reflecting the bias of its writer towards the stereochemical theory.
Little can be gained by further discussion at this juncture, until all the relevant data have been collected, and Dr. Wright has had the opportunity to present his detailed view of the situation from the
standpoint of the vibrational theory, in the next episode of this continuing debate.

ACKNOWLEDGMENTS

I am very grateful to Prof. G. Palmieri, Dr. E. Wanke and Miss S. Vallerga of the University
of Genoa for running the tests with the PAPA machine, to Prof. M.S. Blum of the University of
Georgia for permission to use unpublished pheromone data, and to Mrs. Delpha Venstrom at this
Laboratory for preparing molecular silhouettes. Most emphatically, I record my appreciation to
Dr. Robert H. Wright of the British Columbia Research Council for his cordial agreement to par-
ticipate in this comparison of theories, and for making available to me a considerable amount of
infrared spectral data prior to publication.

REFERENCES

Amoore, J.E. 1952. The stereochemical specificities of human olfactory receptors. *Perfum. essent. Oil Rec. 43,* 321-323, 330.

Amoore, J.E. 1964. Current status of the steric theory of odor. *Ann. N. Y. Acad. Sci. 116,* 457-476.

Amoore, J.E. 1965. Psychophysics of odor. Cold Spring Harb. Symp. quant. Biol. 30, 623-637.

Amoore, J.E. 1967. Specific anosmia : a clue to the olfactory code. *Nature, Lond. 214,* 1095-1098.

Amoore, J.E. 1970. Molecular Basis of Odor. Charles C. Thomas, Springfield, III. 200 p.

Amoore, J.E. and Venstrom, D. 1966. Sensory analysis of odor qualities in terms of the stereochemical theory. *J. Food Sci. 31,* 118-128.

Amoore, J.E. and Venstrom, D. 1967. Correlations between stereochemical assessments and organoleptic analysis of odorous compounds. In "Olfaction and Taste" II, p. 3-17, ed. Hayashi, T. Pergamon Press, Oxford.

Amoore, J.E., Palmieri, G. and Wanke, E. 1967. Molecular shape and odour : pattern analysis by PAPA. *Nature, Lond. 216,* 1084-1087.

Amoore, J.E., Venstrom, D. and Davis, A.R. 1968. Measurement of specific anosmia. *Percept. Mot. Skills 26,* 143-164.

Amoore, J.E., Palmieri, G., Wanke, E. and Blum, M.S. 1969. Ant alarm pheromone activity : correlation with molecular shape by scanning computer. *Science, N.Y. 165,* 1266-1269.

Blum, M.S., Warter, S.L. and Traynham, J.G. 1966. Chemical releasers of social behaviour - VI. The relation of structure to activity of ketones as releasers of alarm for *Iridomyrmex pruinosus* (Roger) *J. Insect Physiol. 12,* 419-427.

Klouwen, M.H. and Ruys, A.H. 1963. Constitution chimique et odeur - III. Comparaison entre les propriétés olfactives de quelques benzaldéhydes substitués et celles des nitrobenzènes substitués de la même manière. *Parfum. Cosmét. Savons 6,* 6-12.

Palmieri, G. and Wanke, E. 1968. A pattern recognition machine. *Kybernetik 4,* (3), 69-80.

Wright, R.H. 1954 a. Odor and chemical constitution. *Nature, Lond. 173,* 831.

Wright, R.H. 1954 b. Odor and molecular vibration. I. Quantum and thermodynamic considerations. *J. appl. Chem., Lond. 4,* 611-615.

Wright, R.H. 1964 a. The Science of Smell. George Allen & Unwin, London, 164 p.

Wright, R.H. 1964 b. Odor and molecular vibration : the far infrared spectra of some perfume chemicals. *Ann. N. Y. Acad. Sci. 116,* 552-558.

Wright, R.H., Huey, R. and Michels, K.M. 1964. Evaluation of far infrared relations to odor by a standard similarity method. *Ann. N. Y. Acad. Sci. 116,* 535-551.

Wright, R.H. and Robson, A. 1969. Basis of odor specificity : homologues of benzaldehyde and nitrobenzene. *Nature, Lond. 222,* 290-292.

DISCUSSION

Chauncy (Chairman) : In your experiments did you allow for the fact that the organic, sweaty acids as they touch the mucus are partially dissociated in water?

Amoore : I do not know the pH of the mucus nor whether the nose detects free acids or ions.

Wright : Since this is explicitly a direct, quantitative comparison of the stereochemical and vibrational theories of odor, I feel it necessary to make the following comments.

The parameter, $\triangle$, which Dr. Amoore uses to compare the profile of a molecule with the profile of some other reference molecule depends upon the location chosen for the molecular centers (or sometimes upon the location of some functional or polar group) and also upon the relative orientations, but what is most important, many very different shapes can have the same value as $\triangle$, as Figure 1 shows.

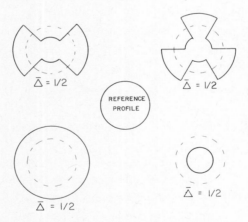

Fig. 1. Relationship between molecular shape and $\triangle$.

This is because in the evaluation of $\triangle$, the sign of δr is lost in the process of averaging $\delta r/r$.

The use of a computer does nothing to improve the situation. It brings to the comparison of the two theories the same degree of impartiality that a man blind from birth would bring to the comparison of two paintings. It can also be questioned how far it is legitimate or useful to make an abnormal condition (that is, anosmia) a major discriminant. Next, in drawing his various regression lines, Dr. Amoore always ignores the reference compound. This seems curious because it is in fact the most statistically certain point of them all and should be given **more** weight than any of the others.

The comparison of our theories should not be limited to only three basic sets of olfactory data. I set no limit. Our latest achievement in the matter of the olfactory threshold is clear-cut and quantitative and presents a most exciting challenge to all other theories.

A very delicate test of the vibrational hypothesis is provided by insect-attractancy studies by the U.S. Department of Agriculture.

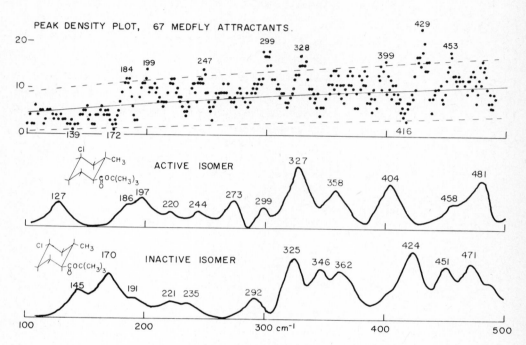

Fig. 2. Attractants for the Mediterranean fruit fly.

At the top of Figure 2 we have a plot based on the spectra of 67 substances which attract the Mediterranean fruit fly, showing the number of compounds having a peak which falls within a band 7 cm^{-1} wide. The solid curve is the number one would expect on the basis of an average of approximately 500 spectra, and the dotted lines are separated by two standard deviations from the expected value. Attractancy for the Medfly appears to be associated with the presence of peaks in certain parts of the spectrum, e.g., at 184 or 247 cm^{-1}, and even more with the absence of peaks in certain other parts, e.g., 176 cm^{-1}. Two stereoisomeric forms of the compound "Trimedlure" have been prepared by McGovern and Beroza (1966) which differ only in the orientation of a chlorine atom, one being biologically active and the other not. A comparison of the spectra shows many points of correspondence between the positions of the peaks in the active isomer and the peak density plot, and corresponding anti-correlations for the inactive isomer.

Fig. 3. Trimedlure Isomers :
Upper : Attractive Isomer,
Lower : Non-attractive Isomer.

Space-filling models of the two isomers shown in Figure 3 provide no such detailed insight and would have little or no predictive value.

Finally, an empirical correlation of olfactory character with any molecular parameter, whether it be molecular shape, or reactivity, or vibrational properties, can not properly be called a **theory of smell**. A theory does much more than simply correlate two things. A correlation is only the beginning of an intellectual process whose object is to bring if possible every aspect of the matter into a clear relation with the whole body of physical and biological science. The stereochemical hypothesis seems to fall far short of this ideal.

Steinbrecht : Why were fatty acid molecules shown with a straight chain?

Amoore : I have made a comparison based on the thermodynamically most likely configurations of a molecule in free space, which is in the straight zigzag chain. All I can say is that it does give a statistically very significant correlation.

Ruzicka : If a keto group instead of a double bond is placed in ring D of the four stereoisomeric androstenols, their male hormone activity is in direct relation with their olfactive behavior. How can you explain that?

Schneider : In reply to Professor Ruzicka, as biologists, we are not puzzled that much, because a hormone receptor could be something like an odor receptor. We are well aware that there are receptors or acceptors for these hormones and they may very well work in a similar way. What does Dr. Amoore think about the fact that some molecules appear very similar but smell differently, while others look very different and smell alike?

Amoore :   I have been comparing each molecule with a standard molecule. Now we shall have to postulate a much more definite receptor site. The progress I envisage is the identification of more primary odors and trying to extract the receptor protein; whereupon someone might do a crystallographic study on the structure of that receptor protein which should hold the answer.

Kovats :   These papers should refer to "correlations" and not to "theories".

Reiner :   I suggest a multifactorial correlation, i.e. a correlation taking into account a steric factor, an active group factor, a partition coefficient factor, etc. That has been done by Hentsch in his study about topography and stereochemistry of enzyme reactive sites, specially on acetylcholine-esterase.

ODOR AND MOLECULAR STRUCTURE

R. Teranishi
Western Utilization Research and Development Division
Agricultural Research Service, U.S. Department of Agriculture
Albany, California 94710

INTRODUCTION

In 1957, a symposium was held here in Geneva on "Molecular Structure and Organoleptic Quality", organized by the Overseas Section of the Society of Chemical Industry. Papers were given by Stoll, Kalmus, Sfiras and Demilliers, Naves, Beets, Wright, Thompson, and a summary by Professor Ruzicka. The relation between molecular structure and odor was perplexing then. It is now. Most of the problems which the mentioned, illustrious men faced, we still face. However, careful chemical work with accompanying sensory testing has brought out some interesting results.

The chemistry of musks has been discussed in detail by Beets (1957), Naves (1957), Stoll (1965), Theimer et al. (1967), and others. It is striking that a wide range of types of compounds exhibit this odor (E.T. Theimer et al. 1967), see fig. 1 : steroids, macrocyclic ketones, nitrocyclohexanes, indanes, tetrahydronaphthalenes, acetophenones, etc. However, even though these com-

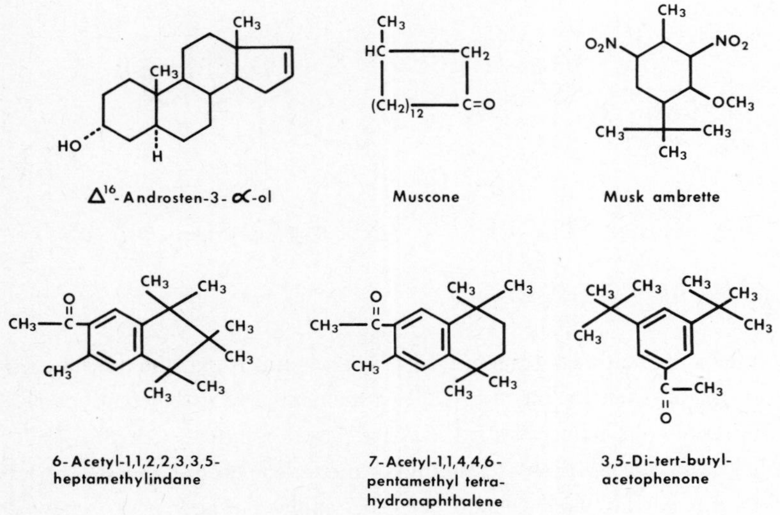

Δ^{16}-Androsten-3-α-ol   Muscone   Musk ambrette

6-Acetyl-1,1,2,2,3,3,5-heptamethylindane

7-Acetyl-1,1,4,4,6-pentamethyl tetrahydronaphthalene

3,5-Di-tert-butyl-acetophenone

Some Musk-Like Compounds.

Fig. 1.

pounds can be placed in this category, Stoll (1965) has pointed out that expert perfumers can distinguish each. As mentioned, much of this work has been discussed; however, I firmly believe that this area must be mentioned again, for this type of elegant chemistry is the basis for our thinking in the area of odor and molecular structure. In fact, it is this type of chemistry which changed the thinking of organic chemists from the two dimensional to the three dimensional world, the world of stereochemistry.

CHANGES IN POTENCY

Beets (1957) and Naves (1957) have quoted Prelog and Ruzicka (1944) and Prelog, Ruzicka, and Wieland (1945). Prelog and co-workers found that the Δ^{16}-androsten-3α-ol epimer has a much stronger musk-like odor than the 3β-epimer, which is judged by some observers as being odorless. Moreover, by reversal of the configuration at C_5, the two epimeric Δ^{16}-aetiocholen-3-ols were obtained, and both are found to be practically odorless. These structures are shown in fig. 2.

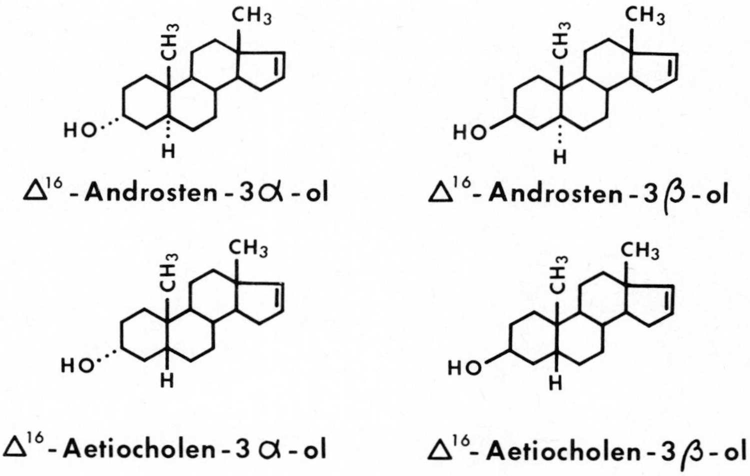

Δ^{16} - Androsten - 3α - ol Δ^{16}- Androsten - 3β - ol

Δ^{16} - Aetiocholen - 3α - ol Δ^{16}- Aetiocholen - 3β - ol

Fig. 2

One of the interesting features of the musk-like compounds is that the odor intensity changes with small changes in structure, but the quality remains such that the compounds still fall in the category of musk-like. Theimer et al. (1967) has shown, see fig. 3, in the series called the "ortho" musks that the addition of a methyl or methylene group can alter the intensity. This observation is not startling since we have seen that epimers of androstenols have different intensities; however, this series is another example of a change in intensity without a great change in quality. No generality seems obvious except that the additional methyl group in the saturated ring seems to give a more potent odor compound.

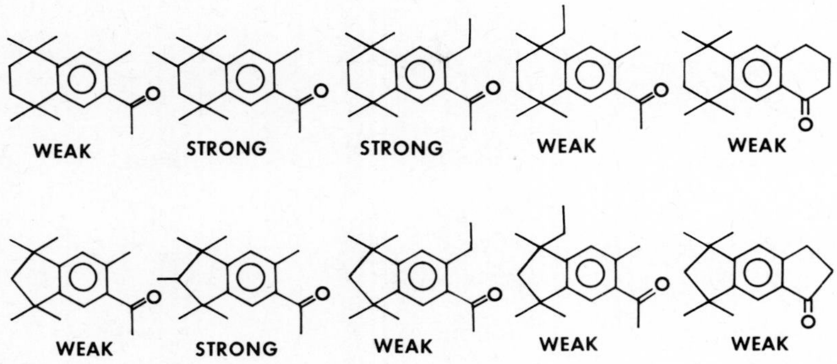

Fig. 3

Stoll (1965) has shown that small differences in internal ketals can change the compound from having a very powerful amber odor to very little or none, see fig. 4. If the two oxygens are present in this type of molecule in a lactone form, the molecules exhibit no odor. Cyclic ethers have strong odors, but the enol ether does not.

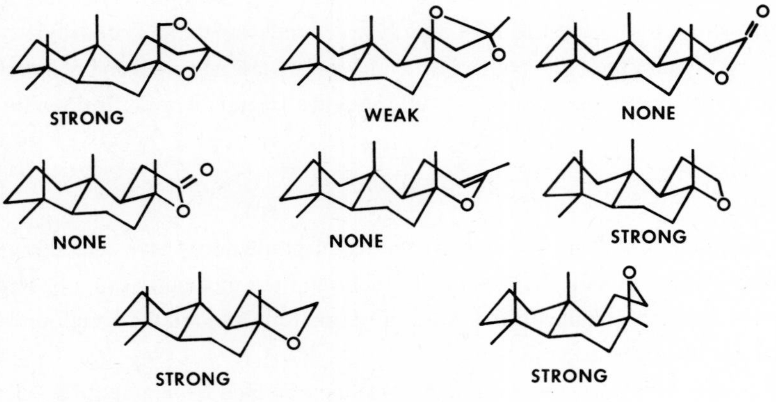

Fig. 4

Interest in wood-type odors has caused the study of alkyl-substituted cyclohexanols (Byers, Jr. 1947; Demole 1964, 1969; Erman 1964; Dahill Jr. et al. 1970) see fig. 5. Stoll (1965) has reported on the comparison of terpenyl-3-cyclohexanols synthesized by Demole (1964). It is not surprising that positional isomers do not exhibit similar odor qualities. As in the case of the

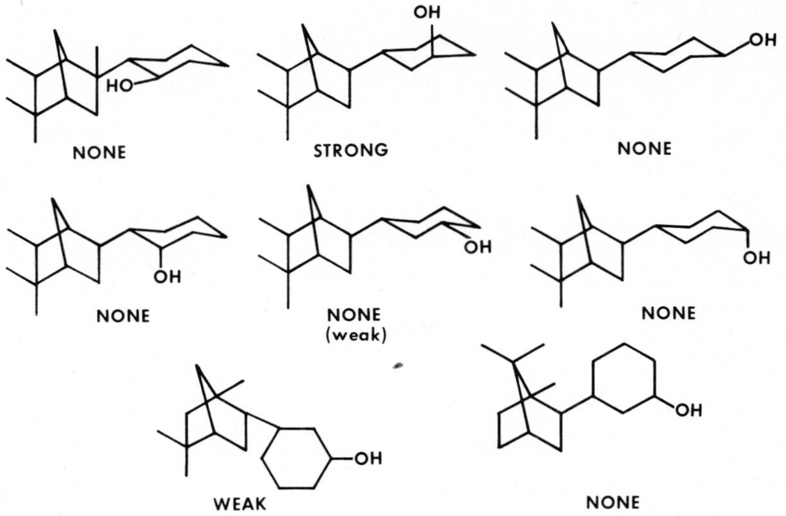

Fig. 5

androstenols (Prelog et al. 1944, 1945), only one epimer has a strong odor; i.e., only the axial hydroxyl compound has an odor, whereas the equatorial hydroxyl compound has little or none.

Demole (1969) has synthesized, by stereospecific reaction, 11 logical isomers out of the theoretically possible 88 terpenyl-cyclohexanols. The others can be rejected since the exo/exo and equatorial isocamphyl are the favored confirmations. The isomers were evaluated by professional perfumers, who verified that the axial 3-(2,2,exo-3-trimethyl-exo-5-norbornyl)-cyclohexanols, threo and erythro, are the only isomers which have the very powerful and genuine sandalwood odor. The other isomers were found to be 20-100 times less fragrant or practically odorless.

CHANGES IN QUALITY

Some compounds with musk-, amber-, and woody-type odors have been discussed very briefly, and in these series, a small change in the structure in the molecules caused a large change in potency but not much in quality. In the series of nootkatone and related compounds, quality changes, but there is little change in potency.

Stevens et al. (1970) have compared nootkatone obtained from grapefruit oil with racemic nootkatone synthesized by Pesaro, Bozzato, and Schudel (1968), and also with other closely related compounds, see fig. 6. The odor thresholds of nootkatone, recrystallized 5 times from grapefruit and the racemic nootkatone, were found to be identical. This finding indicates that there is no difference in the odor of enantiomers. However, this is negative evidence, and more experimental evidence is needed for definite conclusions. In synthesizing nootkatone, Pesaro et al. (1968)

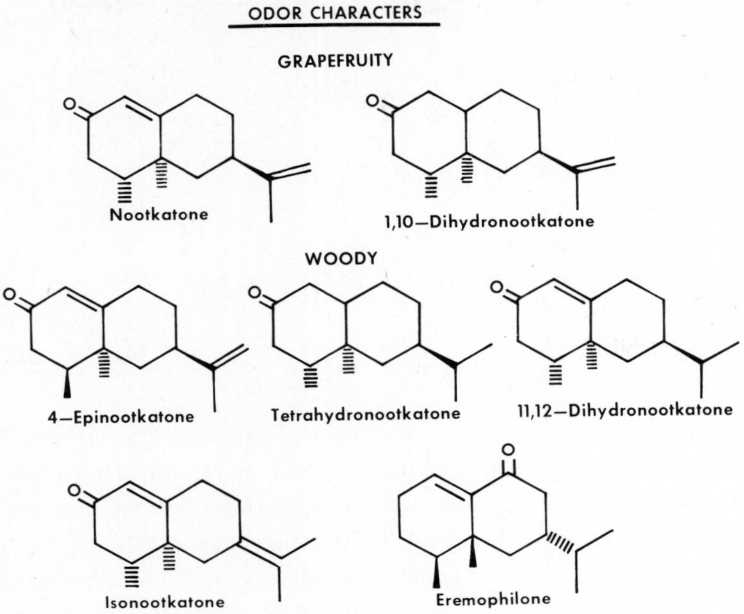

ODOR CHARACTERS

GRAPEFRUITY

Nootkatone   1,10—Dihydronootkatone

WOODY

4—Epinootkatone   Tetrahydronootkatone   11,12—Dihydronootkatone

Isonootkatone   Eremophilone

Fig. 6

made 4-epinootkatone. We thank Dr. P. Schudel, Givaudan-Esrolko, Ltd., Dubendorf-Zurich, Switzerland, for generously sending us samples of racemic nootkatone and 4-epinootkatone. The odor threshold of 4-epinootkatone was found to be slightly lower than that of nootkatone, but the odor qualities of the two epimers are such that they are easily distinguishable. The 4-epinootkatone odor is described as more woody, whereas the nootkatone odor is characteristic of grapefruit. We can easily argue, however, that the grapefruit odor also falls in the woody odor category since nootkatone was originally found in the heartwood of Alaskan yellow cedar by Erdtman and Hirose (1962). The important point is that the epimers have different qualities with little difference in potencies.

Stevens et al. (1970) hydrogenated nootkatone and isolated the two dihydro- and the tetrahydronootkatone. Although it is easily distinguishable from nootkatone, 1,10-dihydronootkatone resembles grapefruit odor the closest by panel judgements. The tetrahydro- and 11,12-dihydronootkatone were considered more woody than 4-epinootkatone. Isonootkatone was obtained by acid treatment of nootkatone, and eremophilone was isolated from the wood oil of *Eremophila mitchelli,* which was kindly sent to us by Dr. J.L. Willis, Museum of Applied Arts and Sciences, Sydney, Australia. These latter compounds, isonootkatone and eremophilone, were considered to be least like nootkatone.

Table 1 shows the odor thresholds of these compounds. At most, there is an order of magnitude in differences in the thresholds. There are no dramatic differences in odor potencies; there is no disappearance of odor.

Table 1. Odor potencies of nootkatone and related compounds

Compound	Thresholds ppb in water
Nootkatone	175
1,10-Dihydronootkatone	50
4-Epinootkatone	100
Tetrahydronootkatone	45
11,12-Dihydronootkatone	500
Isonootkatone	175
Eremophilone	45

Table 2 shows some odor thresholds of compounds with varying potencies. There is a range of about 8 orders of magnitude. Compared to this spread, we can say that the spread of the odor thresholds in the nootkatone-related series is small. Also, this table illustrates the difference between a very characteristic odor, as that of nootkatone which we associate only with grapefruit but with a rather high threshold, and that of a general ester odor, as that of n-amyl acetate with a lower threshold. Also, it was rather surprising to find that the thresholds of humulene and myrcene, compounds

Table 2. Odor thresholds of various compounds

Odorant	μg/liter of water
Ethanol	100,000
Butyric acid	240
Nootkatone	170
Humulene	160
Myrcene	15
n-Amyl acetate	5
n-Decanal	0.1
α- and β-Sinensal	0.05
Methyl mercaptan	0.02
β-Ionone	0.007
2-Methoxy-3-isobutylpyrazine	0.002

D.G. Guadagni, private communication, 1970.

found in many essential oils and generally regarded as not having much odor, are the same or lower than that of nootkatone. The spread of thresholds also illustrates the difficulty of obtaining sensory pure fractions. Less than a tenth of a percent of β-ionone in a nootkatone sample would alter the sensory property of the sample, but such a small amount would not be easily detected spectrometrically.

CHANGES IN POTENCY AND QUALITY

In several series of compounds there are not much quantitative data published, but nevertheless, important data are presented as to the relation of structural change and odors.

Winter (1961) has thoroughly studied the homologs and analogs of 1-(p-hydroxyphenyl)-3-butanone, "raspberry ketone", which was first found in the volatiles of raspberries by Schinz and Seidel (1961). Only 1-(p-hydroxyphenyl)-3-butanone exhibited the strong taste and odor of raspberry, while the methyl ether and acetate derivatives exhibited weak taste and odor of raspberry. All positional isomers, such as the hydroxyl in the ortho- and meta-positions on the phenyl ring and the carbonyl in the 1-, 2-, and 4-positions on the butyl side chain, gave only faint or no raspberry odor or taste. Some of the compounds with one or more carbons added to the butyl side chain gave only a faint odor and taste of raspberry, and others did not exhibit even a fruity odor or taste. This systematic study of the sensory properties of the structural homologs and analogs of the raspberry ketone shows that the structure of 1-(p-hydroxyphenyl)-3-butanone is highly specific with respect to the raspberry flavor, and that even small structural changes diminish the raspberry flavor. It is interesting to note that the acetate of 1-(p-hydroxyphenyl)-3-butanone has been reported as a potent attractant for the male melon fly (Doolittle et al. 1968). The attractiveness of the compound did not change with deuterium substitution, whereas other changes in the structure drastically changed the attractiveness.

Lactones have been known to be important in various characteristic odors for a long time. Moncrieff (1951) has summarized the work of Rothstein (1935) on the preparation and description of some α- and γ-lactones. Of particular interest are the γ-lactones, see fig. 7-1, in which as the side chain is enlarged, the odor changes from coconut to peach to peach-musk. The C_6 side chain lactone has been found in peaches by Sevenants and Jennings (1966). The α-hydroxy-β-methyl-γ-carboxy-$\Delta^{\alpha-\beta}$-γ-hexenolactone, fig. 7-2, is a lactone from a protein hydrolysate (Sulser, personal communication), and this lactone is reported to have a very strong odor and taste of beef bouillon by Sulser et al. (1967). Phthalides and related compounds have also been of interest for a long time. Sedanolide, see fig. 8-5, was isolated from celery oil in 1897 by Ciamician and Silber (1897). Gold and Wilson (1963 a, b) investigated the volatiles from celery and found that 3-isobutyliden-3a, 4-dihydrophthalide, fig. 8-1, 3-isovalidene-3a, 4-dihydrophthalide, fig. 8-2, 3-isobutylidene phthalide, fig. 8-3, and 3-isovalidene phthalide, fig. 8-4, exhibit celery-like odors at levels of 0.1 parts per million in water. Thus, it can be readily seen that a range of quality of odors can be achieved with lactones.

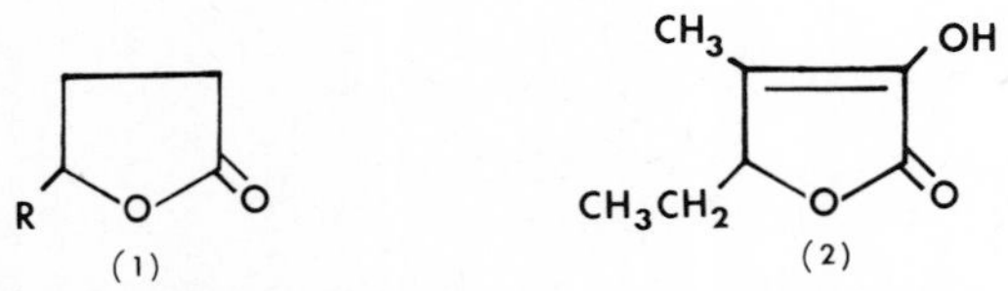

(1)

R = C_5H_{11}, COCONUT
C_6H_{13}, PEACH
C_7H_{15}, PEACH
C_8H_{17}, PEACH-MUSK

(2)

BEEF BOULLION

Fig. 7

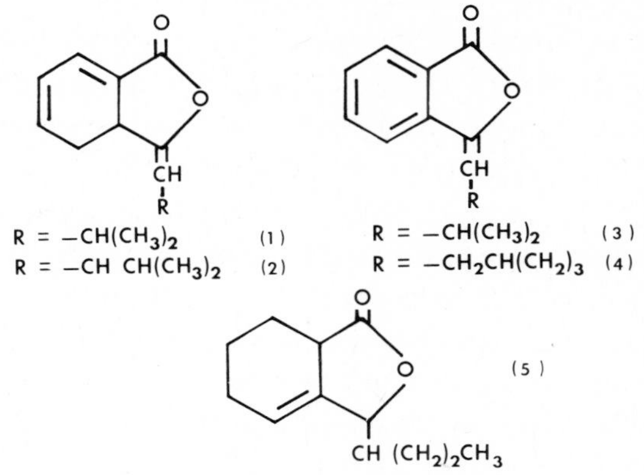

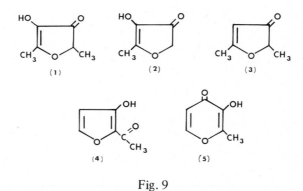

Fig. 8

Some furanones have been reported to have very interesting odors. Hodge et al. (1960, 1963) and Silverstein (1967) have reported that 4-hydroxy-2,5-dimethyl-3(2H)-furanone, fig. 9-1, has a caramel and/or burnt pineapple odor. The 4-hydroxy-5-methyl-3(2H)-furanone, fig. 9-2, is reported to have a roasted chicory root odor by Tonsbeck et al. (1968). Both these compounds are reported as components contributing to beef broth flavor (Tonsbeck et al. 1968). Rosenkranz et al. (1963) have reported that 2,5-dimethyl-3(2H)-furanone has an odor of freshly baked bread, fig. 9-3. Hodge (1967) lists some of the carbohydrate caramelization and pyrolysis products of which isomaltol, fig. 9-4, and maltol, fig. 9-5, must be mentioned as having burnt, pungent, fruity and fragrant, caramel odors, respectively. Thus, there is no doubt that these compounds play an important role in the odor of heated foods, and small differences in the structure of furanones and related compounds do change the quality of odor of such compounds.

Fig. 9

Buttery et al. (1969 a, b) have isolated, identified, and synthesized 2-methoxy-3-isobutyl-pyrazine, the characteristic, highly potent odor component of green bell peppers. The odor threshold of this compound was found to be 2 parts per 10^{12} parts of water. Months after this compound was synthesized by Buttery and co-workers (1969 a, b), his end of the building smelled of freshly chopped green bell peppers. Because of this unusually high odor potency and specific characteristic odor, other pyrazines were investigated (Seifert 1970).

Table 3. Odor thresholds of some pyrazines

Compound	per 10^{12} parts of water
2-Methoxy-3-hexylpyrazine	1
2-Methoxy-3-isobutylpyrazine	2
2-Methoxy-3-propylpyrazine	6
2-Methoxy-3-isopropyl	2
2-Methoxy-3-ethylpyrazine	400
2-Methoxy-3-methylpyrazine	4,000
2-Methoxypyrazine	700,000
2-Isobutylpyrazine	400,000
2,5-Dimethylpyrazine	1,800,000
Pyrazine	175,000,000

Seifert et al. (1970).

Table 3 summarizes the findings of Seifert et al. (1970). The potencies range 10^8. This range is as great as that shown in Table 2, which contains compounds of many different functional groups. If the methoxy- or the alkyl-group is missing, as in 2-methoxypyrazine and with 2-isobutyl- or 2,5-dimethylpyrazines, we see that the potency falls by 10^5 to 10^6. We see that the potency falls by 10^8 if we strip the molecule to the parent molecule, pyrazine.

The most potent compound, 2-methoxy-3-hexylpyrazine, was found to have a bell pepper-like odor, but the one found in the volatiles of pepper, 2-methoxy-3-isobutylpyrazine, was judged to be the most pepper-like. The 2-methoxypropylpyrazine was found to be pepper-like also. The 2-methoxy-3-isopropylpyrazine was judged to be potato-like, but 2-methoxy-3-ethylpyrazine was described as more raw potato-like than the isopropyl compound. The 2-methoxy-3-methylpyrazine was reported as having an odor characteristic of roasted peanuts. The odors of the less potent compounds were found to be not very characteristic.

Pyrazines have been found in roasted peanuts (Mason et al. 1966), cocoa (Marion et al. 1967; Flament et al. 1967; Rizzi 1967), coffee (Bondarovich et al. 1967; Goldman et al. 1967), etc. Such work, plus that by Buttery, shows that pyrazines are important in odors of foods and drinks. The work by Buttery et al. (1969 a, b; Seifert et al. 1970) shows that some pyrazines are very potent and points out that there must be much concentration in order to obtain enough material to identify such compounds. Buttery has shown that potency can change dramatically with relatively small changes in structure.

The smallest difference in molecules, of course, is the difference between enantiomers. Naves (1957) has pointed out that in no case is one enantiomer odorous and the other not. Recently, several investigators have started to re-investigate enantiomers and their odor properties. With the recent advances in separation techniques, we hope that the doubt as to sensory purity will be removed and this problem as to whether enantiomers do or do not have different odors will be resolved.

CONCLUSION

We started this discussion with the study of the androstenols. This type of chemistry has laid the foundation for our thinking in the study of odor and chemical structure. We have discussed how small changes in chemical structure of musk-, amber-, and wood-type odor compounds change the potency greatly but not much in quality. We have looked at nootkatone and related compounds and noticed that quality changes, but the potency does not vary much. We have looked at various cases in which potency and quality change, especially as in the case of the pyrazines. Any generality as to correlations between odor and chemical structure must explain why small differences in structure cause changes in potency and not in quality in some cases, changes in quality but not in potency in other cases, and changes in quality and in potency in still other cases.

REFERENCES

Beets, M.G.J. 1957. Structure and Odor. In "Molecular Structure and Organoleptic Quality", S.C.I. Monograph No. 1, Society of Chemical Industry, London, pp. 54-90.

Bondarovich, H.A., Friedel, P., Krampl, V., Renner, J.A., Shephard, F.W. and Gianturco, M.A. 1967. Volatile Constituents of Coffee. Pyrazines and Other Compounds. *J.Agr.Food Chem. 15,* 1093-1099.

Buttery, R.G., Seifert, R.M., Lundin, R.E., Guadagni, D.G. and Ling, L.C. 1969 a. Characterization of an important aroma component of bell peppers. *Chem. and Ind.* 490-491.

Buttery, R.G., Seifert, R.M., Guadagni, D.G. and Ling, L.C. 1969 b. Characterization of Some Volatile Constituents of Bell Peppers. *J. Agr. Food Chem. 17,* 1322-1327.

Byers, J.R., Jr. 1947. New Woody-Type Odors. *Amer. Perfumer and Essential Oil Rev. 49,* 483-484.

Ciamician, G. and Silber, P. 1897. Ueber die hochsiedenden Bestandteile des Sellerieöls. *Ber. 30,* 492-501, 501-506, 1419-1424, 1424-1427, 1427-1433.

Dahill, R.T., Jr., Dorsky, J. and Easter, W. 1970. A Stereospecific Synthesis of *trans*-3-(*exo*-5-*exo*-Isocamphyl)cyclohexanol. *J. Org. Chem. 35,* 251-253.

Demole, E. 1964. Sur la structure des terpénylphénols obtenus en condensant le camphène avec le phénol, et celle de leurs dérivés hydrogénés à odeur boisée. *Helv. Chim. Acta 47,* 319-338.

Demole, E. 1969. Synthèses stéréospécifiques et étude organoleptique comparée de onze triméthyl-2,2,3-*exo*-norbornyl-5-*exo*-cyclohexanols racémiques. *Helv. Chim. Acta 52,* 2065-2085.

Doolittle, R.E., Beroza, M., Kaiser, I. and Schneider, E.L. 1968. Deuteration of the melon fly attractant, Cue-lure, and its effect on olfactory response and infra-red absorption. *J. Insect Physiol. 14,* 1697-1712.

Erdtman, H. and Hirose, Y. 1962. The Chemistry of the Natural Order Cupressales. The Structure of Nootkatone. *Acta Chem. Scand. 16*, 1311-1314.

Erman, W.F. 1964. The Condensation of Camphene and Phenol. Product Formation via a Direct 2,6-Hydride Transfer. *J. Amer. Chem. Soc. 86*, 2887-2897.

Flament, I., Willhalm, B. and Stoll, M. 1967. Recherches sur les arômes. Sur l'arôme du cacao III. *Helv. Chim. Acta 50*, 2233-2243.

Gold, H.J. and Wilson, C.W. 1963 a. Alkylidene Phthalides and Dihydrophthalides from Celery. *J. Org. Chem. 28*, 985-987.

Gold, H.J. and Wilson, C.W. 1963 b. The Volatile Flavor Substances of Celery. *J. Food Sci. 28*, 484-488.

Goldman, I.M., Seibl, J., Flament, I., Gautschi, F., Winter, M., Willhalm, B. and Stoll, M. 1967. Recherches sur les arômes. Sur l'arôme de café. II. Pyrazines et pyridines. *Helv. Chim. Acta 50*, 694-705.

Hodge, J.E. 1960. Novel Reductones and Methods of Making Them. U.S. Patent 2,936,308.

Hodge, J.E., Fisher, B.E. and Nelson, E.C. 1963. Dicarbonyls, Reductones, and Heterocyclics Produced by Reactions of Reducing Sugars with Secondary Amine Salts. *Am. Soc. Brewing Chemists Proc.* 84-92.

Hodge, J.E. 1967. Origin of Flavor in Foods. Nonenzymatic Browning Reactions. In "Symposium on Foods : The Chemistry and Physiology of Flavors", H.W. Schultz, E.A. Day and L.M. Libbey, Editors, The Avi Publishing Co., Westport, Conn., 465-491.

Marion, J.P., Muggler-Chavan, F., Viani, R., Bricout, J., Reymond, D. and Egli, R.H. 1967. Sur la composition de l'arôme de cacao. *Helv. Chim. Acta 50*, 1509-1522.

Mason, M.E., Johnson, B. and Hamming, M. 1966. Flavor Components of Roasted Peanuts. Some Low Molecular Weight Pyrazine and a Pyrrole. *J. Agr. Food Chem. 14*, 454-460.

Moncrieff, R.W. 1951. "The Chemical Senses". Leonard Hill, Ltd., London, p. 254.

Naves, Y.-R. 1957. The Relationship Between the Stereochemistry and Odorous Properties of Organic Substances. In "Molecular Structure and Organoleptic Quality", S.C.I. Monograph No. 1, Society of Chemical Industry, London, pp. 38-53.

Pesaro, M., Bozzato, G. and Schudel, P. 1968. The Total Synthesis of Racemic Nootkatone. *Chem. Comm.* 1152-1154.

Prelog, V. and Ruzicka, L. 1944. Untersuchungen über Organextrakte. Ueber zwei moschusartig riechende Steroide aus Schweinetestes-Extrakten. *Helv. Chim. Acta 27*, 61-71.

Prelog, V., Ruzicka, L., Meister, P. and Wieland, P. 1945. Steroide und Sexualhormone. Untersuchungen über den Zusammenhang zwischen Konstitution und Geruch bei Steroiden. *Helv. Chim. Acta 28*, 618-627.

Rizzi, G.P. 1967. The Occurence of Simple Alkylpyrazines in Cocoa Butter. *J. Agr. Food Chem. 15*, 549-551.

Rosenkranz, R.E., Allner, K., Good, R., Philipsborn, W.V. and Eugster, C.H. 1963. Zur Kenntnis der Chemie einfacher Furenidone (β-Hydroxyfurane). *Helv. Chim. Acta 46*, 1259-1285.

Rothstein, B. 1935. Odeur et constitution. Sur quelques γ-butyrolactones-α-substituées. *Bull. Chim. Soc. 2*, 80-90. Odeur et constitution γ-butyrolactones-γ-substituées, ibid., 1936-1944.

Schinz, H. and Seidel, D.F. 1961. Nachtrag zu der Arbeit Nr. 194 von H. Schinz and C.F. Seidel in *Helv. 40,* 1839 (1957). *Helv. Chim. Acta 44,* 278.

Seifert, R.M., Buttery, R.G., Guadagni, D.G., Black, D.R. and Harris, J.G. 1970. Synthesis of Some 2-Methoxy-3-Alkylpyrazines with Strong Bell Pepper-Like Odors. *J. Agr. Food Chem. 18,* 246-249.

Sevenants, M.R. and Jennings, W.G. 1966. Volatile Components of Peach. *J. Food Sci. 31,* 81-86.

Silverstein, R.M. 1967. Pineapple Flavor. In "Symposium on Foods : The Chemistry and Physiology of Flavors", H.W. Schultz, E.A. Day, and L.M. Libbey, Editors, The Avi Publishing Co., Westport, Conn., 450-461.

Stevens, K.L., Guadagni, D.G. and Stern, D.J. 1970. Odor Character and Threshold Values of Nootkatone and Related Compounds. *J. Sci. Food Agr.,* submitted.

Stoll, M. 1965. De l'Effet Important de Différences Chimiques Minimes sur la Perception de l'Odorat. *Revue de Laryngologie.* G. Portmann, Bordeaux, pp. 972-981.

Sulser, H., Depizzol, J. and Buchi, W. 1967. A Probable Flavoring Principle in Vegetable-Protein Hydrolysates. *J. Food Sci. 32,* 611-615.

Theimer, E.T. and Davies, J.T. 1967. Olfaction, Musk Odor, and Molecular Properties. *J. Agr. Food Chem. 15,* 6-14.

Tonsbeck, C.H., Plancken, A.J. and Weerdhof, T.V.D. 1968. Components Contributing to Beef Flavor. Isolation of 4-Hydroxy-5-methyl-3(2H)-furanone and its 2,5-Dimethyl Homolog from Beef Broth. *J. Agr. Food Chem. 16,* 1016-1021.

Winter, M. 1961. Odeur et constitution. Sur des homologues et analogues de la p-hydroxyphenyl-1-butanone-3 (cétone de framboise). *Helv. Chim. Acta 44, 2110-2121.*

DISCUSSION

Randebrock : Concerning the qualitative data of your potency measurements of odorants diluted with air and water respectively, did you take into account the water solubility on the threshold values of the substances, and the vapor pressure on the intensity curves?

Teranishi : The threshold data are given in terms of concentration in water solution. Of course, the important fact is the number of molecules reaching the olfactory nerves. To know what the vapor pressure is above the solutions, we must know the partition coefficients. Buttery et al. (1969, 1970) have obtained some vapor pressure measurements. There is a gradual increase in volatility for the higher molecular weight homologs with alcohols, ketones, esters, and aldehydes for the range at least of C_1 to C_{10}. Therefore, if the vapor pressures above an aqueous solution were considered, the thresholds of the pyrazines would look even more impressive. We did consider solubilities of the substances and the vapor pressures above aqueous solutions. However, because most fruit and vegetables are predominantly water, because of the difficulties involved in obtaining partition coefficients and vapor pressure data for compounds of extremely low thresholds, and because there have been some relationships established between vapor pressure and concentration, we present our thresholds in terms of concentration in water.

Harper :   You always refer to threshold as a definition of potency, but in many cases this is not so.

Teranishi :   There is an amount needed for detection, and this we define as the threshold of detection. There is an amount needed to recognize the material, and this we define as the threshold of recognition. Moreover, as the concentration is increased, the quality may change. We use the threshold of detection as the first order approximation of the potency.

Dravnieks :   The speaker is to be complimented for presenting the excellent review and examples on relation of the molecular structure and odor. It should be added, however, that although some changes in the molecule, e.g., a shift of methyl group to another position, appear unimpressive when one inspects the respective structural formulae, these changes may have significant impact on simple intermolecular interaction of the odorant with condensed phases. Thus, a shift of a methyl group from a position next to a secondary hydroxyl group to a position one methylene link further changes the Kovats Index of the substance in Carbowax 20M by 100 units — the substance is more "polar" since the -OH group is more accessible for intermolecular interactions.

REFERENCES

Buttery, R.G., Ling, L.C. and Guadagni, D.G. 1969. Food Volatiles : Volatilities of aldehydes, ketones, and esters in dilute water solution. *J. Agric. Food Chem. 17,* 385-389.

Buttery, R.G., Bomben, J.L., Guadagni, D.G., and Ling, L.C. 1970. Some considerations of the volatiles of flavor compounds in foods. 160th A.C.S. National Meeting, Chicago, Ill., September 14-18.

RELATIONSHIP BETWEEN ODOR SENSATION AND STEREOCHEMISTRY

OF DECALIN RING COMPOUNDS

G. Ohloff

Research Laboratories, Firmenich et Cie.,
1211 Geneva 8, Switzerland.

Ambergris odor is composed of six distinguishable qualities (Ohloff, 1969). We have found that the characteristic ambergris odor or any of its individual component qualities arise exclusively in compounds having a decalin ring system of a strictly determined stereochemistry symbolised by formula A.

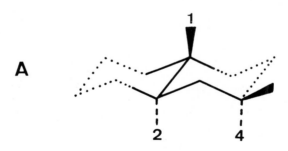

The importance of this relationship will be illustrated by a small number of examples.

In the dihydroambrinol (**1**) series (Ohloff and Giersch, 1970 a) only the diastereomer, **1a**, possesses four of the characteristic odor qualities of ambergris (wet earthy-musty, sea water-seaweed, tobacco, and sandalwood). The interchange of the methyl group and the angular hydrogen as shown in the compound **2a** (Ohloff et al., 1970 b) causes a reduction of the odor qualities of the sandalwood note. Both *trans*-decalin derivatives **3a** and **3b**, in which the oxygen function occupies the angular position, have the typical odor of musty wet earth (Marshall and Hochstetler, 1968). The animal note of musc is even produced by polycyclic ring compounds incorporating the *trans*-decalol system as shown in the well known example of Δ^{16}-androsten-3a-ol (Prelog et al., 1945).

By variation of these four structures (formulas **1 – 4**) in a certain manner virtually all six odor qualities which are very close to those of the ambergris notes can be generated. The only requirement which each molecule has to fulfill is the **1, 2, 4-triaxial arrangement of the substituents in the decalin ring system A,** one of these axial groups being an oxygen function.

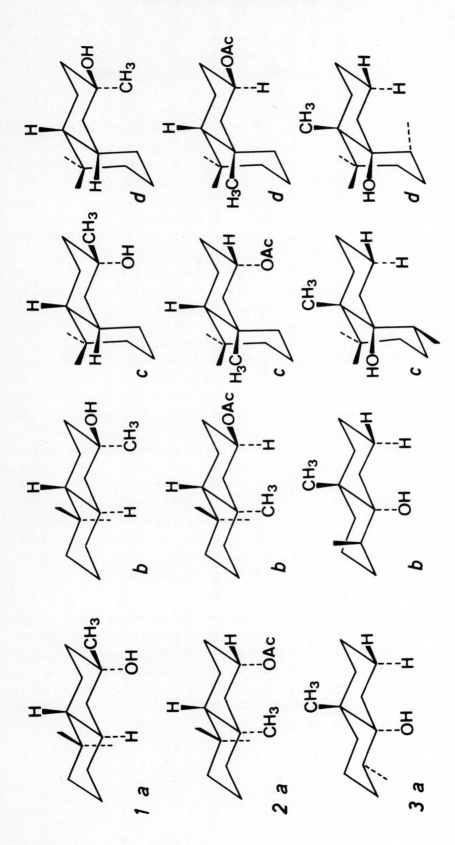

Fig. 1. Diastereomers of dihydroambrinol (1), 1,1,5-trimethyl-7-acetoxy-decalin (2), geosmines (3).

By reversing the position of the oxygen function so that it adopts the equatorial position, as in the cases of **1b**, **2b** and 3β-androstenol, the molecules lose practically all their odor. In components derived from *cis*-decalin (**1c, 1d, 2c, 2d, 3c, 3d**) a camphor-like odor predominates. The two epimeric $\triangle^{16}$-etiocholen-3-ols are practically odorless (Prelog et al., 1945).

Apparently the receptor also accepts that part of a π-orbital which is arranged on the same side as the axial substituent in the 2- or 4-position respectively. This is the case with the double bond in the distinctly ambergris-like odorant, a-ambrinol (Stoll et al., 1956), and with the ketone derived from **2a** and **2b**. a-Ambrinol, with the hydroxyl group in equatorial position (Armour et al., 1959), does not smell at all whereas the ketone with the *cis*-fused decalin ring system (derived from **2c** and **2d**) has a camphor-like odor (Ohloff and Giersch, 1970 a). The position of the secondary methyl group in the case of the naturally occurring earth metabolite geosmin **3** (Marshall and Hochstetler, 1968) has only a negligible effect on the odor quality. Its omission leads to practically the same odor sensation. The situation is again totally different in the *cis*-decalin series, the diastereomers **3c** and **3d**, for example, having a camphor-like odor.

Ambrox (**4a**) and isoambrox (**4b**) (Stoll and Hinder, 1950) as well as the two intramolecular C-18 ketals **5a** and **5b** (Scheidegger et al., 1962) are very characteristic ambergris fragrances. Although the stereochemistry of the oxygen function is the reverse, the odors of the pairs **4a** and **4b** differ only quantitatively but not qualitatively. The reason for this is the triggering of the odor sensation from the two different constellations **4a** and **4b**. In the ambrox case the tetrahydrofuran ring of stereoformula **4a′** simulates one part of the *trans*-decalin ring system whereas the receptor site seems to accept isoambrox only in the form of **4b′**. The 1,2,4-triaxial rule will be followed even from three of the possibilities of the two diastereomeric intramolecular C-18 ketals, namely the forms **5a** and **5b** which are just as active as **5a′**. In the *trans*-decalin ring system of **5a′** as well as in the *cis* case **5b′** an O-heteroatom replaces a carbon atom.

Molecules in which the free rotation is hindered can simulate a ring system behaving like a *trans*-decalin ring system. Apparently this phenomenon comes into play in the case of the iso-camphyl cyclohexanol series. Among the numerous possible isomeric structures only the two compounds **6** and **7** which fulfill the triaxial rule develop a strong sandalwood odor (Demole, 1969).

It would appear that the 1,2,4-triaxial rule can also be applied to the monocyclic ketone **8** which has an odor similar to that of $\triangle^{16}$-androstenone (Beets and Theimer, 1970). According to our present knowledge of conformational analysis, the molecule is folded and therefore the substituents are fixed as represented by the formula **8′**. In this steric arrangement, however, it is easy to recognize the structural elements similar to those of the steroid system.

The conclusion of this investigation is that odor qualities similar to those of ambergris are generated by compounds having a strictly determined stereochemistry. It was found that this phenomenon arose from the 1,2,4-triaxial arrangement of the substituents in a decalin ring system, or in a molecule with an equivalent configuration. The position of the oxygen function within the triaxial system moves the sensation in the direction of different but distinct qualities. On the other hand the surroundings of the center triggering the odor, as well as the chemical nature of the oxygen substituent seems to be of minor importance.

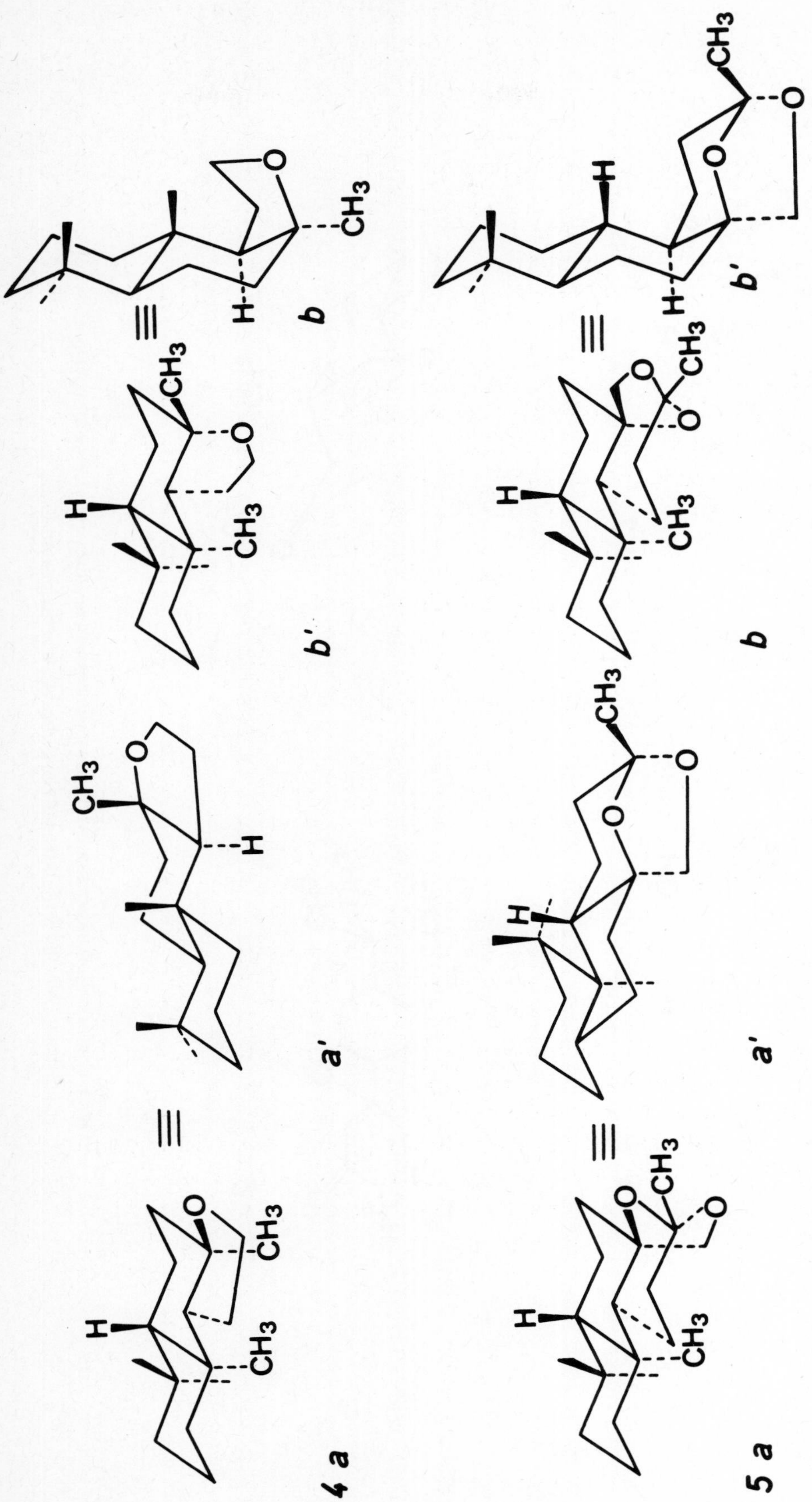

Fig. 2. Diastereomers of 4a ambrox, 4b isoambrox and the intramolecular C-18 ketals 5a and 5b.

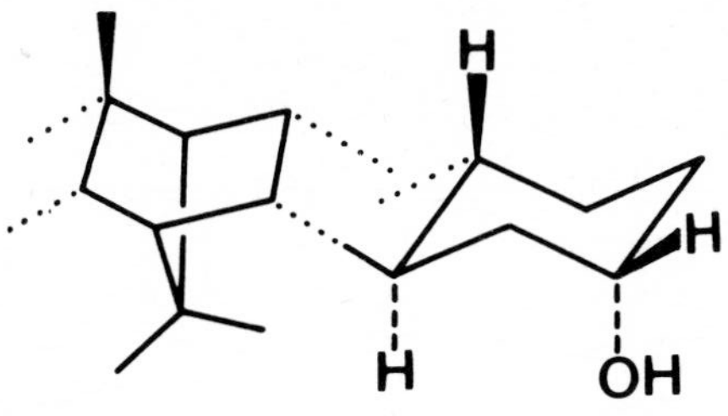

6

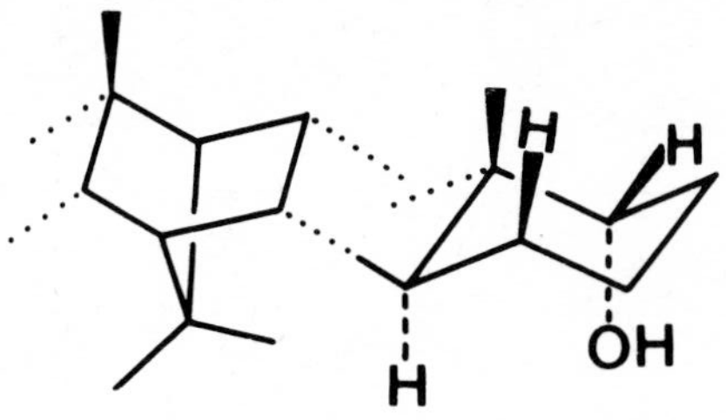

7

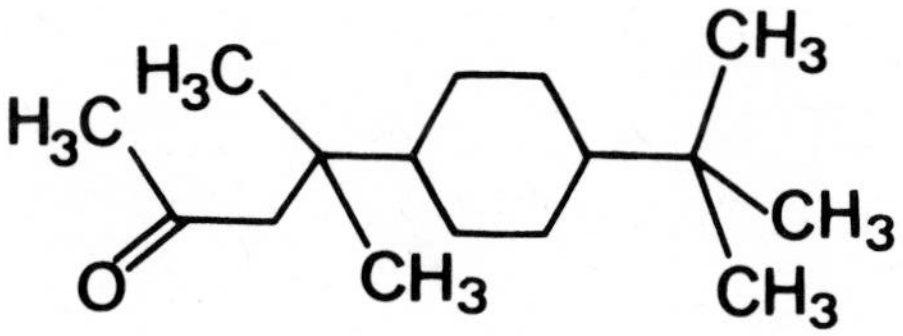

8

≡

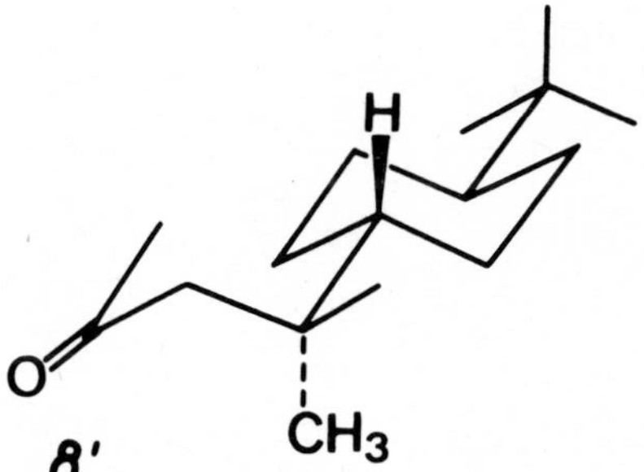

8′

Fig. 3. Conformation of isocamphyl-3-*threo*-cyclohexanol **6a**, isocamphyl-3-*erythro*-cyclohexanol **6b**, and 4-(4′-tert.-butylcyclohexyl)-4-methylpentan-2-one.

REFERENCES

Armour, A.G., Büchi, G., Eschenmoser, A. and Stroni, A. 1959. Synthese und stereochemie der isomeren Ambrinole. *Helv. Chim. Acta 42,* 2233-2244.

Beets, M.G.J. and Theimer, E.T. 1970. Odor similarity between structurally unrelated odorants. Proceedings of the CIBA Foundation Symposium on Taste and Smell in Vertebrates, p. 313-323, J. & A. Churchill, London.

Demole, E. 1969. Synthèses stéréospécifiques et étude organoleptique comparée de onze triméthyl-2,2,3-exo-norbornyl-5-exo-cyclohexanols racémiques. *Helv. Chim. Acta 52,* 2065-2085.

Marshall, J.A. and Hochstetler, R. 1968. The synthesis of (±)-geosmin and other 1.10-dimethyl-9-decalol isomers. *J. Org. Chem. 33,* 2593-2595.

Ohloff, G. 1969. Chemie der Geruchs- und Geschmacksstoffe. *Fortschr. chem. Forsch. 12,* H. 2, p. 185-251, Springer-Verlag, Berlin-Heidelberg-New York.

Ohloff, G. and Giersch, W. 1970 a. Unpublished results.

Ohloff, G., Sundt, E. and Näf, F. 1970 b. Unpublished results.

Prelog, V., Ruzicka, L., Meister, P. and Wieland, P. 1945. Untersuchungen über den Zusammenhang zwischen Konstitution und Geruch bei Steroiden. *Helv. Chim. Acta 28,* 618-627.

Scheidegger, U., Schaffner, K. and Jeger, O. 1962. Ueber die Struktur und Umwandlungen von zwei stereoisomeren Riechstoffen $C_{18}H_{30}O_2$ aus Manool. *Helv. Chim. Acta 45,* 400-435.

Stoll, M. and Hinder, M. 1950. Sur les époxydes hydroaromatiques à odeur ambrée. *Helv. Chim. Acta 33,* 1308-1312.

Stoll, M., Seidel, C.F., Willhalm, B. and Hinder, M. 1956. Sur la constitution de l'ambrinol. *Helv. Chim. Acta 39,* 183-199.

THE SENSORY PROPERTIES OF (+)-NOOTKATONE

AND RELATED COMPOUNDS

G. Ohloff and W. Giersch

Research Laboratories, Firmenich et Cie.,
1211 Geneva 8, Switzerland.

Our investigations on the relationship between odor and structure in the nootkatone series are based on 10 sesquiterpene derivatives. All these compounds were optically active and of related stereochemistry (fig. 1).

We found that those compounds with a **fruity odor** have a **bitter taste** whereas those not smelling fruity have no taste. These sensory properties (fig. 1) are strictly related to the functional group, and the conformation of the oxygen-free part of the molecule contains a double bond in a defined position. The double bond most distant from the carbonyl group is of special importance for the initiation of the fruity odor and bitter taste. 1, 10-Dihydro-nootkatone as well as β, γ-nootkatone have the grapefruity odor most like nootkatone in that series. Concerning the intensity of the grapefruity odor, β, γ-nootkatone is the most powerful substance we know at the present time.

ORIGIN AND PURITY OF THE COMPOUNDS

(+)-**Nootkatone** was isolated from grapefruit oil (Mc Leod, 1965) and recrystallized several times from pentane. m.p. 36 - 37°.

(+)-**4-epi-Nootkatone** is described in the unpublished synthetic work of Dr. H.C. Barrett (M.I.T.). We are grateful to Prof. G. Büchi for furnishing the pure sample.

(–)-**iso-Nootkatone and (–)-4-epi-iso-nootkatone** result from the reaction of 4-chloro-2-pentanone and (-)-*cis*-m-menth-8-en-6-one (Ohloff and Giersch, 1968) in the presence of sodium hydride and tetrahydrofuran at reflux in an argon atmosphere. Separation and purification of the two sesquiterpene ketones was by chromatography on a silica gel column using a solvent mixture of benzene-ethyl acetate (9 : 1).

(–)-β, γ-**Nootkatone.** (+)-Nootkatone in t-butanol solution was treated with excess of potassium t-butoxide for 18 hr at room temperature. The deconjugation occurs in practically quantitative yield.

(+)-**Nardostachone** was made from (+)-nootkatone (Pinder, 1970).

(–)-β-**Vetivone** was isolated from the carbonyl fraction of vetiver oil by successive treatment with Girard reagent followed by careful chromatography on silica gel. This procedure was executed by Dr. Fracheboud in our laboratory.

(+)-α-**Vetivone** was obtained by heating (+)-nootkatone with 20-fold amount of aqueous phosphoric acid (H_3PO_4, 10 %) at 100° during 54 hr. The equilibrium mixture contained 79 % (+)-α-veti-

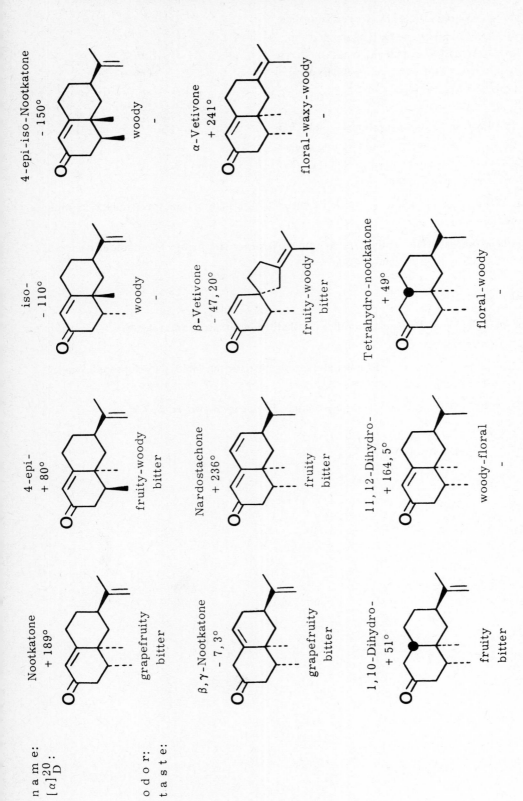

Fig. 1. Structural, physical and sensory properties of (+)–nootkatone and related compounds.

vone, 14 % (+)–nootkatone and 7 % of an isomer ([a]$_D$ + 235.9°), and on the basis of spectral evidence had an endocyclic double bond at C-6 (**1**). The sensory properties of the endocyclic isomer are similar to those of (+)–*a*-vetivone.

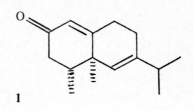

1

(+)–**1,10-Dihydronootkatone** is obtained pure from the partial hydrogenation of (+)-nootkatone in absolute ethanol over palladium on charcoal in the presence of 2 % potassium hydroxide.

(+)–**11, 12-Dihydronootkatone** was made by partial hydrogenation using tris-(triphenylphosphine)chlororhodium (Pinder, 1970).

(+)–**Tetrahydronootkatone** resulted from complete hydrogenation of (+)-nootkatone over palladium on charcoal in ethyl acetate solution.

REFERENCES

McLeod, W.D., jr. 1965. The constitution of nootkatone, nootkatene and valencene. *Tetrahedron Letters,* 4779-4783.

Ohloff, G. and Giersch, W. 1968. Säurekatalysierte Isomerisierung von α, β-Cyclopropyloxiranen. *Helv. Chim. Acta 51,* 1328-1342.

Pinder, A.R. 1970. The structure of Nardostachone. *Tetrahedron Letters,* 413-415.

GUSTATORY, BEHAVIORAL AND PHARMACOLOGICAL MANIFESTATIONS OF CHEMORECEPTION IN MAN

Roland Fischer

Department of Psychiatry, Division of Behavioral and Neurobiological Sciences,
and Department of Pharmacology, The College of Medicine of the Ohio State University,
Columbus, Ohio U.S.A. 43210.

INTRODUCTION

You will agree with Freud (1954) that a decisive step in phylogenetic development occurred when the **Urmensch** first got his nose above the female genitalia and assumed an erect position and a bipedal gait. The hands were thus freed to pre-process food, and in this sense became part of the gustatory and digestive system, while the new point-of-view of the eyes "bade him behold the sky and, upright, lift his gaze unto the stars." (Ovid, from Straus, 1965). Indeed, only now was he "Zum Sehen Geboren, zum Schauen bestellt" (Goethe). The supreme importance of vision thus relegated olfaction to an auxiliary but nevertheless decisive role in the perceptual-behavioral interpretation of central nervous system activities in the bedroom, bathroom and kitchen. Of course, for those of us who still walk on four or more legs the sense of smell remains the **sens extraordinaire** (see, for example, Michael 1970; Murphy, 1970; Morrison, 1970).

In the following we will deal with gustatory chemoreception and with some of those aspects of smell which are inseparable from this topic. Beginning with the molecular level, we will proceed to the highest, the systemic or organismic level of complexity, at which man, the self-referential system, perceptually-behaviorally interprets the change induced in his central nervous system activity by chemical stimuli.

Because our first paper on taste was published in 1959 (Fischer and Griffin, 1959), and since then our experimental and theoretical approaches have been refined and varied extensively, it is appropriate to describe our evolved methodology at this point.

THE DETERMINATION OF TASTE THRESHOLDS

Materials

Taste thresholds are determined in a room with a reasonably constant temperature of, preferably, 22 C^0.* Compounds for tasting, such as quinine sulfate, should be of pharmacopeia or first-grade reagent quality, water is glass-still double-distilled or de-ionized, and paper cups (75-100 ml) are used for tasting.

PREPARATION OF TASTE SOLUTIONS, USING QUININE AS AN EXAMPLE

Weigh 0.2886 g quinine sulfate $(C_{20}H_{24}N_2O_2)_2.H_2SO_4.2H_2O$ into a light-protected volumetric flask, and fill to one liter with distilled water. Using a magnetic stirrer, dissolve the

---

*The temperature of 22^0C is that at which the lowest threshold for any compound will be obtained. (Griffin, 1966; Fischer and Kaelbing, 1967).

quinine, raising the temperature to not more than 40°C while stirring. The flask now contains Solution No. 11, or 7.50×10^{-4} M quinine sulfate which is a saturated solution at room temperature (see Table 1). (A subject who cannot taste quinine in solution below this concentration is said to have a taste threshold of 11).

Dilute 500 ml of this Solution No. 11 to 1000 ml with distilled water. Mix, pour 500 ml into a dark brown bottle, and label "Quinine Solution No. 10." Dilute the remaining 500 ml of Solution No. 10 to 1000 ml with distilled water, mix and label "Quinine Solution No. 9." By thus halving each previous concentration, prepare Solutions 8, 7, 6, 5, 4, 3, 2, 1 and 0, the last being 3.66×10^{-7} M quinine sulfate, the most dilute solution generally needed (see Table 1).

The taste-testing method is a double-blind, forced-choice procedure (Fischer and Griffin, 1961 a, 1964 a), based on the design of Sir Ronald Fisher (1951). For comparative tables listing solution numbers (of macro taste thresholds) with corresponding molarity and the nearest corresponding solution numbers used by Harris and Kalmus (1949) and Barnicot, Harris and Kalmus (1951) for quinine, phenylthiourea and 6-n-propylthiouracil (PROP), see Fischer and Griffin (1964 a). The macro threshold solution number 14 is thus 2^{10} or 1024 times the concentration of Solution No. 4. Taste thresholds should be determined at least twice for each subject, since half a population displays increased taste sensitivity (lowering by one threshold) on retest. The variability of the macro quinine threshold is $\pm$ 1 solution number, but stress, such as lack of sleep, etc., may result in decreased sensitivity of up to two thresholds. Exposure to repeated macro taste threshold determinations does not result in practice-induced lowering of taste thresholds.

Determination of macro taste thresholds

1) Pour about 5 ml each of Solutions No. 2, 4 and 6 into three different small paper cups which have been previously marked in such a way that they can only be seen by inverting the cups. Into three other identical but **unmarked** cups pour about 5 ml **placebo**, i.e., distilled water of exactly the same temperature as the (quinine) solutions. A constant-temperature water-bath, as in fig. 1, is helpful (Griffin, 1966).

Present to the subject Solution No. 2 and a cup of placebo, making sure that before and after tasting each solution or placebo, he rinses his mouth with distilled water from a large cup or, as in fig. 1, from a bent glass tube inserted into a 250 ml bottle of distilled water in the water bath. The subject is instructed not to swallow solution, placebo or rinse, but to merely "turn them over" in his mouth, and is asked if he can clearly taste a **difference** between the two cups of solution and placebo. If he cannot, repeat with Solution No. 4 and placebo and, if he cannot taste this, with Solution No. 6, and so on. When the subject can taste a difference, repeat with the next lower solution. For example, if he reports a difference between Solution No. 4 and placebo, repeat the above procedure with Solution No. 3. The subject's approximate quinine threshold is now known, and the tester can proceed to measure the subject's **exact** macro quinine taste threshold.

2) If the subject can differentiate Solution No. 3 and placebo, but not the previous Solution No. 2, he is presented with four invisibly-marked cups each containing about 5 ml of Solution No. 3, and four unmarked cups each containing about 5 ml placebo. This is done in a **double-blind** manner, i.e., neither the tester nor the subject knows at this time the contents of the eight cups which the tester randomly distributes in a circle. The cups rest on the water sur-

face in the water bath (cf. fig. 1). The subject is now required to sort the cups into solutions and placebos, always rinsing his mouth with distilled water before and after tasting. The accuracy of the subject's identification of solution and placebo is unknown to both tester and subject until each series of eight cups is sorted into two groups of four. If the subject has correctly sorted the two groups of cups at Solution No. 3, he repeats the procedure at half the preceding concentration, i.e., with Solution No. 2. The subject's macro threshold will be 3 if he makes two mistakes in sorting Solution No. 2, while one mistake in sorting necessitates a "repetition"; if the subject makes any mistakes at all in this "repetition", his taste **threshold** is regarded as the next higher **correctly sorted** quinine solution, (Solution No. 3 in the example). Correctly sorting the "repetition" of Solution No. 2 calls for moving down one more solution number.

Fig. 1. Taste-testing unit as used in our laboratory since 1961 for determination of taste thresholds and jnd values at controlled temperature ($\pm 0.1^\circ C$).

PREPARATION OF TASTE SOLUTIONS OTHER THAN QUININE :

6-n-Propylthiouracil (PROP)

This odorless compound is preferred to phenylthiourea (phenylthiocarbamide =PTC) for genetic studies of tasting and non-tasting. Solution No. 14 is the highest concentration; dissolve 1.0212 g in distilled water to one liter. Solution No. 13 is one-half this concentration, etc.

Hydrochloric Acid (HCl)

Highest concentration for tasting is 0.012 M, equivalent to Solution No. 15; prepare by diluting 60 ml 0.2 N HCl to one liter with distilled water. Solution No. 14 is half this concentration, etc.

Sodium Saccharinate

Highest concentration for tasting is Solution No. 15; dissolve 2.4621 g in distilled water to one liter. Solution No. 14 is half this concentration, etc.

Taste thresholds in a population for quinine will range from Solution No. 0 to No.11, and for "tasters" of PROP, from Solution No. 1 to 9, or from 10-14 for the "non-tasters". For HCl, thresholds range from 6-14, for NaCl, 10-19 and, for sucrose, from 10-20. The differences in threshold range for the various compounds are mainly due to differences in their water solubility.

Determination of semimicro taste thresholds

The procedure for determination of semimicro taste thresholds, initially called by Fischer and Griffin (1964 a) "fine"and "tenfold-resolved" threshold determination, is analogous to that of the macro procedure except that each macro interval is exponentially subdivided into ten subunits. Since a macro interval represents a $3.66 \times 10^{-7} \times 2^1$ concentration increase, each semimicro interval will be a $3.66 \times 10^{-7} \times 2^{1/10}$ increment. A table of taste solutions in semimicro intervals expressed in molarity of fractional solution number can be found in Fischer and Griffin (1964 a).

Micro procedure for the determination of taste threshold

In the micro procedure, macro taste thresholds are **linearly** subdivided into **fifty** subunits, while the semimicro procedure resolves them **exponentially** tenfold. The chief advantage of linearly interpolating macro intervals is its economy in time. In mathematical terms the concentration intervals in the micro procedure correspond to 1 per cent of the next higher macro interval of 2^n, where n stands for the "solution number" and is related to the molarity by the expression : molarity = $3.66 \times 10^{-7} \times 2^n$.

During a determination, the micro threshold is approached in 10 per cent decrements of the macro threshold until the lowest 10 percent decrement which can be tasted is found. The micro threshold is then approached in smaller intervals of 5, 2 and finally 1 per cent decrements until the lowest micro step which can be correctly sorted is found. This concentration, then, represents the micro threshold of the subject.*

In order to obtain reproducible results the subject is required to **practive** taste testing near the threshold concentration one day prior to an experiment. The chemical structure of the compound used for practice is unimportant.

---

* The experimental error in molar concentration of any test solution is estimated to be not greater than $0.004 \times 3.66 \times 10^{-7} \times 2^n$ where n represents a solution number (Table 1).

CONCENTRATIONS OF MACRO THRESHOLDS IN MOLARITY

Macro threshold in Solution Numbers	Concentration in Molarity
21	$7.68 \pm 0.03^* \times 10^{-1}$
20	$3.84 \pm 0.015 \times 10^{-1}$
19	$1.92 \pm 0.01 \times 10^{-1}$
18	$9.60 \pm 0.04 \times 10^{-2}$
17	$4.80 \pm 0.02 \times 10^{-2}$
16	$2.40 \pm 0.01 \times 10^{-2}$
15	$1.20 \pm 0.005 \times 10^{-2}$
14	$6.00 \pm 0.02 \times 10^{-3}$
13	$3.00 \pm 0.01 \times 10^{-3}$
12	$1.50 \pm 0.01 \times 10^{-3}$
11	$7.50 \pm 0.03 \times 10^{-4}$
10	$3.75 \pm 0.015 \times 10^{-4}$
9	$1.875 \pm 0.01 \times 10^{-4}$
8	$9.375 \pm 0.04 \times 10^{-5}$
7	$4.69 \pm 0.02 \times 10^{-5}$
6	$2.34 \pm 0.01 \times 10^{-5}$
5	$1.17 \pm 0.005 \times 10^{-5}$
4	$5.86 \pm 0.02 \times 10^{-6}$
3	$2.93 \pm 0.01 \times 10^{-6}$
2	$1.465 \pm 0.06 \times 10^{-6}$
1	$7.32 \pm 0.03 \times 10^{-7}$
0	$3.66 \pm 0.015 \times 10^{-7}$

*Estimated standard error.

Table 1

MICRODETERMINATION OF A JUST-NOTICEABLE TASTE DIFFERENCE (jnd)
(Fischer and Kaelbling, 1967)

By using comparison and reference concentrations as test solutions, the smallest concentration difference – the just-noticeable difference – can be distinguished. The reference concentration is the lower limit of a jnd and is held constant while the comparison concentration is decreased in micro steps until the upper limit of a jnd is found. A jnd is conveniently expressed as a Weber ratio, i.e., a dimensionless ratio $\frac{\Delta S}{S}$, in per cent, where ΔS denotes the concentration difference between the lower and upper limits of a jnd in taste and S denotes the lower limit.

Example of a microdetermination of a jnd. We first determine whether the subject can taste the compound (e.g. sodium saccharinate) in a concentration corresponding to that of Solution No. 8 (see Table 1) using four cups of distilled water and four cups of solution No. 8. Most subjects ($\approx$ 85 percent of a population are able to sort out solution No. 8 from distilled water, when No. 8 concentration is their reference or baseline. For the minority, the next higher concentration, corresponding to Solution No. 9, is used as the reference or baseline.

The same procedure is then repeated, but with four cups of a sweet versus four cups of a "more sweet" solution, that is, the determination of a jnd proceeds using the reference concentration as the lower limit of the jnd. The initial comparison solution given to the subject is approximately 40 percent more concentrated than the reference.

Four cups containing 5 ml each of the reference solution and four cups containing 5 ml each of the comparison solution are then randomized. The subject tastes each solution, rinsing his mouth before and after tasting with distilled water of the same temperature as the solutions. After he correctly sorts the first set of eight cups, he is given the next consisting of four reference (threshold) concentrations and four comparisons. The concentration of the comparison solutions is reduced by 1 percent of Solution No. 8 to 70 per cent of 8, i.e., 6.5630×10^{-5} M. When the subject correctly sorts this set, the concentration of the comparison solutions is lowered to 69 per cent of 8. If the subject does not correctly sort this set, the upper limit of the jnd is 70 per cent of 8. If a subject "misses" a cup, he is told to try once more and only if he misses again is he offered another series of the next higher concentration. Thus we expose each subject to forced-choice with series of ascending or descending concentrations. It has been ascertained that the size of the jnd is the same whether obtained by ascending or descending concentrations.

Each subject is told at the beginning of the test that if he is undecided about a particular cup, he may put that cup aside but must make a decision later on in the test.

EVALUATION OF TEST RESULTS

The first macro threshold determination may be for some subjects a new, anxiety-provoking experience, in which case only a retest under more familiar conditions will result in a reliable threshold value (see Fig. 2) . An individual's second threshold will not change significantly on further testing except under conditions of stress or to a small extent as a result of seasonal variation (Fischer, 1967 a). Moreover, repeated **macro** taste threshold determinations do not result in practice-induced lowering of thresholds (learning), whereas repeated (2-3 times per week) testing with the **semimicro** and **micro** methods does. (Fischer and Griffin, 1964 a; Fischer et al. 1965). On the other hand, acutely anxious patients under identical experimental conditions show a Gaussian distribution of quinine thresholds with a mean threshold of 7. (Fischer et al. 1965). Another group of subjects, outpatients with migrain headache was studied

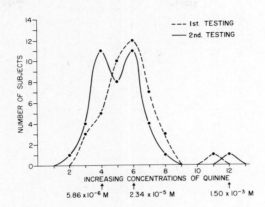

Fig. 2. Gaussian distributions of quinine macro taste thresholds of 41 college-age volunteers, repeatedly determined within two weeks. 50 per cent of the population is variable, i.e., lowers its taste threshold, and
thus shifts the mean threshold from 6 to 5 on retest.

nine years ago by Dr. William Hunt and myself, with results of retesting shown in Fig. 3 for
quinine and Fig. 4 for PROP.

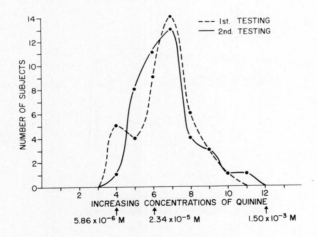

Fig. 3. Gaussian distributions of quinine taste thresholds
of 29 outpatients with migrain headache, determined within two weeks. The mean macro threshold is 7 on both tests, corresponding to a fourfold higher concentration than the mean quinine
threshold, for a college-age population on retest.

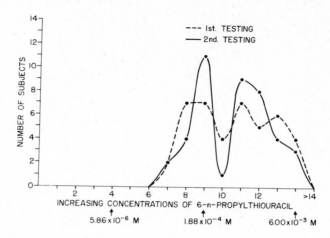

Fig. 4. Bimodal distribution of 6-n-propylthiouracil (PROP)
macro taste thresholds for 29 outpatients with migrain headache, determined within two weeks. The
antimode at PROP 10 appears only on retest. In
comparison, college-age students display an antimode for PROP at 9 for both tests. (Fischer and Griffin, 1964 a).

A related problem is the reduced attention span of certain categories of subjects, e.g. acutely
anxious mental patients (Fig. 5). Their mean quinine threshold value was stable through even

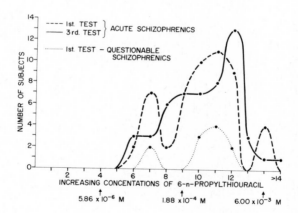

Fig. 5. Distribution of taste sensitivity for PROP
in 48 acutely ill female schizophrenics and
11 questionable schizophrenics. The patients
were tested six times; only the first and
third tests (on the first and sixth days of
treatment with trifluoperazine) are plotted.

six retests (Fischer et al. 1965), however, their PROP threshold distribution never displayed a distinct antimode at 9-10. This was most likely due to a decreased attention span coupled with fatigue and confusion, since PROP tests always followed a quinine test.

Data solely based on one test determination should therefore not be taken too seriously, and significantly elevated taste thresholds based on one test in populations with reduced attention span and mental confusion may merely reflect the inability of the patient to perform the test.

TASTE AND DRUG RECEPTORS; SENSITIVE AND INSENSITIVE SUBJECTS

In a plot of the taste thresholds of a population for quinine, sucrose, methylene blue, etc. the two tails of the Gaussian distribution (Fischer et al. 1962; Fischer and Griffin, 1963) represent the particularly **sensitive** subjects with low taste thresholds (left tail in Fig. 6), and the particularly taste **insensitive** subjects with high taste thresholds (on the right tail). The divisions on the abscissa of the figure represent increasingly higher macro taste thresholds from left to right, while the ordinate stands for the number of subjects tested. A very insensitive taster may need a 512 times more concentrated quinine solution than a very sensitive taster in order to differentiate it from distilled water.

We now know that such a difference in sensitivity to a virtually unlimited variety of "Gaussian" compounds is the reflection of a more general, organismic or systemic sensitivity (Fischer and Griffin, 1964 a, and 1963). Systemic sensitivity refers to the reactivity of the human system in a variety of its manifestations. For example, a sensitive taster, i.e., one who needs fewer quinine molecules in water to differentiate the solution from water alone, will also need fewer molecules of a drug to elicit a specific pharmacological effect.

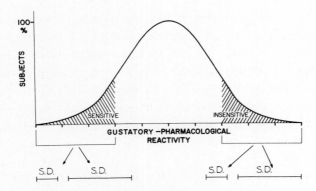

Fig. 6. Gaussian distribution of gustatory and pharmacological sensitivity. Sensitive tasters of quinine (or of any other compound with a normal distribution of taste threshold) are sensitive drug reactors, whereas insensitive tasters need a greater amount of the same drug to induce a comparable effect.

We examined the appearance of Parkinsonian tremor, an "extrapyramidal" side effect appearing after administration of phenothiazine-type tranquilizers in acutely anxious mental patients. Dividing those patients who developed drug-induced tremor into taste sensitives and insensitives, we found that the taste sensitives were also drug sensitive, i.e., they required about two-and-a-half times smaller cumulative doses of the tranquilizer trifluoperazine than the taste-insensitive patients. (Fischer et al. 1965).

Other manifestations of this relation between taste threshold and drug reactivity or threshold were also found. For example, we measured reaction time with the written Serial Seven Test (in which the time for a patient to successively deduct seven from one hundred until zero is reached is measured, with his number of mistakes) on the first, third, sixth, ninth and twelfth days after starting the tranquilization schedule and again prior to dismissal. The taste-sensitive subjects always had significantly faster mean "reaction times" than the insensitives (see Table 2).

That these relations are not peculiar to acute mental patients nor to phenothiazine tranquilizers, was demonstrated by Joyce et al. (1968), who measured the increase in heart rate in 38 healthy medical student volunteers after a standard dose of hyoscine butylbromide. When experimenters with no knowledge of the subjects' autonomic responses determined their taste thresholds for hyoscine butylbromide, they found that taste-sensitives responded with a significantly higher heart-rate increase than insensitives, although everyone was given the same dose of drug. Two other autonomic measures (visual accomodation and salivary secretion) showed the same relationship to taste thresholds, although the comparative tests were separated by an interval of one year. Such a differential reactivity to the same dose of a centrally acting drug was already observed in animals (Hecht et al. 1961). We have called attention to the relation between an individual's taste threshold for a drug and that drug's median lethal dose (LD_{50}*) in laboratory animals. The LD_{50} may, therefore, be justly regarded as the ultimate **systemic** pharmacological response.**

The relation between taste threshold and drug reactivity, therefore, should enable one to estimate the optimal, individualized, therapeutic dose of a drug, instead of mechanically vending drugs by body weight. However, the relation between taste threshold and drug reactivity should not be indiscriminately applied to all drugs. Exceptions might be those non-metabolized, highly lipid-insoluble medicinal agents which are not reabsorbed by the kidneys (and thus pass through into the urine) and whose intensity or pharmacological effect does not vary with mammalian species (Griffin, 1966). It is to these two classes of drugs, those which do and those

---

*   The LD_{50} denotes the amount of drug which is necessary to kill 50 out of 100 animals, and constitutes, in our opinion, another proof of differential drug reactivity, since only 50 of 100 animals are deadly sensitive to a given dose.

**   The relation between two **systemic** phenomena, taste threshold and drug threshold, indicates that the drug thresholds of organs **isolated** from the system (isolated tissue preparations, for example) will not reflect taste thresholds, and indeed this is the case.

bjects	1st day	3rd day	6th day	9th day	12th day	Prior to Dismissal	
1	—	—	300	—	300	300	
2	60	60	45	60	45	—	
3	120	60	75	60	120	50	
4	43	30	40	40	60	33	insensitive
5	60	60	60	30	40	60	tasters
6	120	90	90	80	90	90	N = 10
7	105	105	45	60	50	60	
8	120	120	90	300	60	150	
9	60	45	105	50	90	120	
10	—	90	120	120	120	120	
1	120	60	45	60	90	60	
2	90	90	90	90	90	60	
3	25	60	20	20	60	35	
4	90	60	60	45	45	45	
5	90	60	60	45	90	60	sensitive
6	100	90	90	90	60	90	tasters
7	—	60	90	60	90	90	N = 13
8	60	90	60	60	60	45	
9	120	90	150	45	45	60	
10	30	30	30	60	45	60	
11	90	—	60	30	30	30	
12	60	90	60	45	60	45	
13	—	—	60	60	90	60	

able 2. Both groups of patients, that is, the insensitive tasters, N = 10 (from an original twelve, of whom two could not be tested for reaction time), and the sensitive-tasters, N = 13, are those who (from a total of forty-eight patients, all under the same drug dose schedule) displayed extrapyramidal side effects.

	S.E.	Mean	
insensitive tasters	± 66.1	93.2	n = 129 Students t = 3.407
sensitive tasters	± 25.4	65.3	significance level < 0.001

which probably do not show the taste threshold-drug threshold relation, that Brodie and Reid refer when remarking that,

> " a clinical investigator whose experience has been confined mainly to drugs like quaternary ammonium compounds, thiazide diuretics and polar antibiotics, might suspect that the importance of individual differences has been grossly exaggerated. In contrast, psychiatrists who have been concerned with psychotherapeutic agents and other liposoluble drugs are aware of the wide divergencies in drug response, but may not attribute them to difference in drug metabolism. We know of few liposoluble drugs whose rates of metabolism in man are not highly variable." (Brodie and Reid, 1969)

With this restriction in mind we can now regard an individual's oral cavity as a pharmacological preparation, and his taste threshold as an indicator of pharmacological reactivity (Fischer and Griffin, 1964 a).

In order to be a drug, therefore, a compound must be able to elicit a taste sensation. We define a drug,* (which can also be thought of as a colorless dye,**) as

1) a nitrogenous base, the nitrogen atom of which may be replaced isosterically;
2) water soluble in concentrations of at least 1×10^{-6} M;***
3) bitter-tasting in concentrations of at least 6.0×10^{-3} M.

No exception to this rule has yet been found since its publication. (Fischer and Griffin, 1963).

MODELS OF CHEMORECEPTION

a) General

As the initial step in both taste sensation and organismic drug response the sorption or affinity between molecules and receptors should be considered. In model experiments, I have shown (Fischer et al. 1951; Fischer, 1956) that a compound's affinity,**** i.e., the amount sorbed from its aqueous solution and retained by the fibrous wool protein, α-keratin, or even an ovalbumin-pepsin mixture, ***** is consistently positively correlated with that compound's biological

---

*   This generalization might not include those non-metabolized, highly lipid-insoluble agents which are not reabsorbed by the kidney.

**   This useful concept was originated by Läuger et al. (1944).

***   This minimum water solubility is based on Ferguson's principle, (Ferguson, 1939) which we regard as an extension of Henry's law. This principle states that, as solubility in water decreases (in ascending series of homologs, for example), thermodynamic activity does not increase proportionately, so that eventually a substance is obtained which has no biological activity.

****   Due to the acidic isoelectric point (pH=4.5-4.7) of wool protein, biologically active cationic compounds such as quaternary disinfectants and phenothiazine-type tranquilizers display proportionately greater affinity for α-keratin and consequently greater biological activity than do anionic compounds. Therefore, data on the protein affinity and biological activity of various compounds can only be meaningfully interpreted within each (cationic or anionic) series.

***** A pepsin-ovalbumin solution (0.05 per cent w/v) was used as a model of taste receptor sites, and the affinity of various compounds for this enzyme-substrate solution was spectrophotometrically measured at the compound's λ-maximum.

activity. The greater its affinity for protein, the more biologically potent will the compound
prove to be. We have quantitatively assessed "biological potency" as the ability of a compound
to elicit a bitter taste (Fischer, 1960), as the compound's optimal pharmacologically effective
dose, (Fischer and Griffin, 1964 a) its bactericidal effectiveness (Fischer and Seidenberg, 1951)
and relative toxicity in 14 day-old tadpoles of the South African frog *Xenopus levis*. (Fischer,
1956).

Why should α-keratin be a suitable model of taste and drug chemoreceptors? As Vinnikov
said,

> "**all** the receptor cells of sense organs are provided with specifically differentiated
> motile antennae, possessing two central and nine pairs of peripheral fibrils. The-
> se are made up of myosin-like macromolecular chains". (Vinnikov, 1965)

Myosin, of course, is a protein of the α-keratin type, just as hair and wool are (see also
Fischer and Larose, 1952) and the suggestion has been made that animal receptors evolved from
the α-keratinous flagellae of single-celled, flagellated organisms (Vinnikov, 1965, Filshie and Rogers, 1961). We may even look at α-keratin as a receptor with chiral sites. From an aqueous solution of (±)– mandelic acid, wool, for instance, will selectively adsorb (±)– mandelic acid (Bradley and Easty, 1951), resolution probably occurring at the L-arginine and L-lysine sites, mainly in the crystalline region of the fiber (Bradley and Easty, 1953).

Fig. 7 shows the pleated-sheet structure for scleroproteins proposed by Pauling and Corey,
(1951) and Pauling et al. (1951). Perutz et al. (1964) pointed out that such a structure would

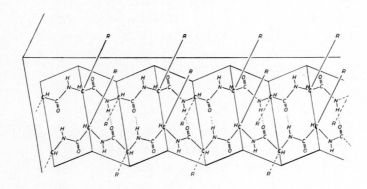

Fig. 7. Pleated-sheet structure of β-keratin according to Pauling and Corey(1951),
reproduced from Fig. 8 in P. Karlson, "Kurzes Lehrbuch der Biochemie",
7th Edition, Georg Thieme Verlag, Stuttgart, 1970, by kind permission of
the publisher. According to Meyer, Mark and Astbury (1961), the polypeptide chains must be almost fully extended in the silk fibroin-β-keratin group
with a fiber period of 6.7-7 Å, whereas, in the α-keratin-myosin-fibrinogen
group, the identity period is 5.1-5.4 Å.

give a strong meridional reflection of 1.5 Å corresponding to the segment of the helix occupied by one residue.* Indeed, this reflection is present in the patterns given by hair, quill, muscle, various other proteins and polypeptides, and constitutes the strongest support for the a-helical structure. An alternative system has also been described, where the axes of the a-helices are straight (Parry, 1969). These terms a- and β-keratin are generally applied to one type of helical (a) and to the extended (β) polypeptide chain conformations in proteins and polypeptides (Bendit, 1966). When wool or hair (a-keratin) is extended by about 50 per cent in water at 20^0C, it changes to β-keratin; on release in water at 20^0C, it reverts to the original a-keratin structure, the two "mechanical stereoisomers" thus being interconverted by slow extension and relaxation in water (Haukaas et al. 1969; Alexander and Hudson, 1954). The cystine cross-linkages of the stretched β-fibers enhance the chemical reactivity of the protein (Speakman, 1936; Wolfram, 1965), as shown by the increased dye sorption ('Gram positiveness') (Fischer and Larose, 1952) of wool which has been strained by knot formation (Fischer and Larose, 1952) or permanently set (Eckert, 1969). The structural changes accompanying knot formation in the wool fiber (bond strain and ultimate breakage of cystine disulphide linkages) may be analogous to the transformation (unfolding, unmasking) in reversibly denatured, excited-stimulated β-keratin-like protein displaying enhanced chemical reactivity in nervous structures (Wolfgram and Rose, 1962).

The reversible changes from Gram-positive to Gram-negative staining behavior in wool and bacterial membranes (Fischer and Larose, 1952 b; Fischer, 1953; Larose and Fischer, 1953) remind us that a-keratin behaves like a mixed ion-exchanger (Stigter, 1964), with regeneration of disulphide bonds during refolding or recovery from the Gram-positive β-keratin to the Gram-negative, a-keratin state (Fischer and Larose, 1952 a), with the a-helix in aqueous solution apparently stabilized by hydrophobic side chain interaction (Lotan et al. 1966). The initial steps in chemoreception might therefore be governed by the affinity of the chemical stimulus for the a-keratin-type receptor, which converts the proteinous moiety of the receptor to a stretched, β-keratin-like structure, while the ensuing desorption of the chemical stimulus by saliva (in taste) may contribute to a relaxation and conversion to the original a-keratin state.**

b) Specific

When various ratios of subthreshold concentrations of PROP and quinine are offered to a subject for taste-testing, certain ratios of these binary mixtures will be tasted, while others will not. The data which underlie such a discontinuous taste pattern (Rubin et al. 1962; Fischer et al. 1965 a) (see Fig. 8) can also be plotted as a drug dose-response curve, with certain combinations of the two drugs producing positive or negative synergism (Fischer and Griffin, 1964 a).

---

*   It is tempting to speculate that the a- to β-keratin conformation as a model of the chemoreception mechanism can be reconciled with such studies as that of Shallenberger and Acree (1967), Kubota and Kubo (1969), and Mazur, Schlatter and Goldkamp (1969), who all attempt to account for correlations between taste sensation and steric configuration of a tasted compound.

**   The most recent alternative to the coiled-coil model has been proposed by Parry (1970). The segmented structure for the a-helical strands in the multi-stranded rope model for a-proteins allows each strand to assume a conformation reminiscent of a coiled coil, except that all the distortion from folding is confined to the part of the protein chain between segments.

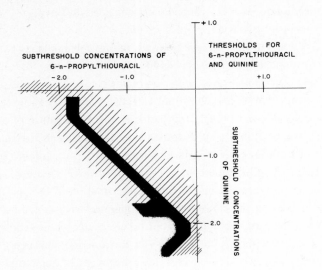

Fig. 8. Tasted (hatched area) and non-tasted (shaded area) binary mixtures of subthreshold so-
    lutions of 6-n-propylthiouracil (PROP) and quinine for a typical insensitive taster of PROP
    Concentration intervals are powers of 2; the intersection of the ordinate and abscissa re-
    presents the taste thresholds of the subject for PROP and quinine alone. Negative numbers,
    therefore, denote subthreshold concentrations of PROP (abscissa) and quinine (ordinate)
    in the binary mixtures. Note that the discontinuity of taste sensation is not proportional
    to the concentration of the components in the binary mixture. For example, the subthres-
    hold concentration – 1.9 PROP and –1.0 quinine is a tasted combination, whereas –1.0
    PROP and –1.2 quinine (in the shaded area) is not tasted. The still lower concentration
    of –1.0 PROP and –1.4 quinine, however, is again a tasted combination.

Two models can account for this discontinuous pattern:

(1) The isothermal taste diagram (Rubin et al. 1962; Fischer et al. 1965 a) is actually a physico-chemical 'translation' of the pharmacological dose-response curve for the same binary mixtures of quinine and PROP illustrated in Fig. 8. The boundary between tasted and non-
tasted combinations defines a phase diagram of the taste phenomenon, and quite closely resembles an isothermal solubility diagram for a pair of salts having an ion in common, both of which
are dissolved in a fixed amount of solvent (Rubin et al. 1962). In order to deal with the taste
phenomenon in terms of an isothermal diagram we determined the boundary (thresholds) theoretically, comparing it to the envelope which separates the experimentally obtained non-tasted
combinations from the tasted ones.

The qualitative agreement between the leaf-like shape of the theoretical threshold
curve (Rubin et al. 1962), and the experimentally obtained envelopes enables us to reason as
follows : at threshold, a certain number of sites in the receptor are occupied by either quinine
molecules or an unknown compound formed from quinine and PROP. The attraction forces between quinine molecules, or between the molecules of the new unknown compound, are greater
than those between molecules of quinine and the new compound. We regard, therefore, the taste threshold of a drug or of a mixture of drugs as a specific state of part of the receptor with a
given number of receptor sites occupied by drug molecules. Taste threshold concentrations of a

subject for various compounds depend then upon the number of available sites as well as the magnitudes of the partition coefficients. Another inference which can be drawn is that the receptor sites, per unit receptor-volume, are probably farther apart in a typical insensitive taster than in a typical sensitive taster of PROP.

(11) Fig. 9 A and 9 B illustrate our second, multiple-site, chemoreceptor model, which approaches the phenomenon of taste threshold as a biological response of the all-or-none type (Gander et al. 1964). The model postulates that biological response occurs only when a certain fraction of the receptor system is converted to active species. Each receptor is assumed to contain multiple binding-sites and a single translating site; the formation of active species proceeds through reversible sorption, first to the binding sites and finally to the translating site. The model, when applied to taste-response data of binary mixtures of quinine and PROP (Fig. 9 A and 9 B), accounts for the discontinuity in the pattern of taste sensation, and illustrates (Fig. 8) that, as the concentration of quinine increases, the fraction of biologically active species decreases. Further increase in the concentration of quinine increases the fraction of active species. This demonstrates the agonistic and antagonistic effect of the same drug in a binary drug mixture, and is in accord with the findings of Van Rossum and Hurkmans (Gander et al. 1964) that the same drug may act both as agonist and antagonist. In our terminology, agonistic (taste) or antagonistic (no taste) response occurs when two or more drugs combine with the receptor to form biologically active or inactive receptor species, respectively.

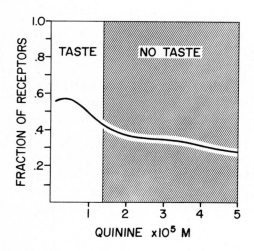 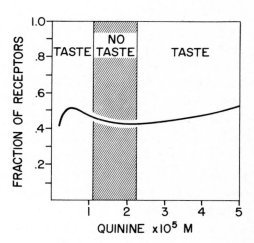

Fig. 9 A and 9 B. Calculated fraction of receptors in biologically active forms compared to taste response for various concentrations of quinine in binary mixtures of subthreshold solutions of quinine and 6-n-propylthiouracil (PROP).

A. The PROP concentration in the binary mixture was held constant at $\underline{0.4 \times 10^{-3} \text{ M}}$, while the quinine was varied from 0.2×10^{-3} to 5×10^{-5} M.

B. The PROP concentration in the binary mixture was held constant at $\underline{1 \times 10^{-3} \text{ M}}$, while the quinine was varied from 0.2×10^{-3} to 5×10^{-5}M.

Our first isothermal solubility diagram model may also explain the **taste of water** resulting from mixtures of subthreshold concentrations of ions. Since sodium, magnesium and calcium ions are significantly more potent stimuli of the subthreshold-to-taste mechanism than is the potassium ion, it appears that the palatability of water is probably a function of its cation content (Cox et al. 1955). And because the taste of water depends upon the subjects's prior adaptation to sodium chloride, individual differences in water palatability may be realted to differences in sodium chloride content in the saliva (Bartoshuk et al. 1964). Incidentally, the sour-bitter taste of water after adaptation to sodium chloride may be regarded as a gustatory "afterimage" analogous to visual afterimages (Bartoshuk et al. 1964).

Both of our specific models may also contribute to the understanding of the taste synergism which results when a compound that has no palatable taste such as glycine is added in a specific ratio to palatable compounds, such as L-glutamic acid or L-aspartic acid, to improve the existing taste (Yokotsuka et al. 1969). The model may even be extended to odor synergism, i.e., when a bouquet results from combinations of (subthreshold) concentrations of odorous compounds which do not have a pleasant smell by themselves.

Our models may also account for the significant lowering of sucrose taste thresholds by sodium chloride present in as little as 0.001 M concentrations (Anderson, 1955). We have found that we can lower the threshold of a subject by a fourfold concentration decrement and at the same time change the taste quality by offering a solution of quinine and sodium chloride, each compound in concentrations two thresholds below the subject's threshold for that compound. The resulting taste, interestingly, is sweet. Not surprisingly, the addition of subthreshold concentrations of sodium chloride to sucrose solutions "makes the sweetness perception 'rounder' " (Jellinek, 1964). These data on sodium chloride in mixtures with other compounds implicate ionic strength as a factor in taste, and for instance, a tenfold increase in the ionic strength of a sodium chloride solution lowers a sodium saccharinate threshold by 15 per cent.

The provocative practices of the "homeopathic" Hahnemann school may be related to this observation. They use very small concentrations of drugs either alone or in mixtures, and hold that in many cases the effects of small doses differ from those of large doses not only quantitatively, but also qualitatively. Some of the dilutions in question are in the range of $1 : 10^6$ by weight, and are thus in the concentration range of the taste solutions which are represented below the shaded area in Fig. 8. It is therefore conceivable that highly dilute, inactive concentrations of drugs (synergistically) become pharmacologically effective when combined in certain ratios with components of body fluids or with other drugs. This line of reasoning may also explain why drugs used in folk medicine are in the form of mixtures and, specifically, why a completely inactive fraction may potentiate another only mildly active fraction (Hijmans, 1962).

PERCEPTUAL-BEHAVIORAL CONCOMITANTS OF TASTE SENSITIVITY

(1) Personality and taste

Drug reactivity is not the only parameter significantly related to taste threshold. We have found that taste-sensitive and thus drug-reactive subjects, as a group (on the left tail of the curve in Fig. 6), are intuitors and, to a lesser extent, introverts on the Myers-Briggs Type Indicator (MBTI), whereas insensitive and thus less drug-reactive subjects are factual down-to-earth sensors and, to a lesser extent, extroverts on the MBTI (Corlis et al. 1967). Additional confirmation of this relation presented itself when we realized that 95 per cent of our volunteers for experiments with hallucinogenic drugs (like LSD, psilocybin and mescaline) proved to be sensitive tasters and MBTI intuitors. This association between **intuition** and **taste sensitivity** also agrees with the observations of McGlothlin and Cohen, who noted a preponderance of intuitors among their student subjects with "positive attitudes toward taking LSD" (McGlothlin and Cohen, 1965).

Evidently, intuitive and, to some extent, introverted subjects are more easily attracted to an intuitive and introverted (drug) experience.

It is interesting to not that, according to MacKinnon, three out of every four people in the United States are sensors, leaving only one quarter intuitors (MacKinnon, 1966). On the other hand, at least nine out of ten of that country's creative writers, mathematicians, research scientists and architects are recruited from this last category (MacKinnon, 1967). Among college students, as are 95 per cent of our volunteers for hallucinogenic drug experimentation, 60 per cent are taste-sensitive intuitors. We wonder whether this intuitive, taste-sensitive and creative segment of the population is proportionately represented in Congress.

(II) Food preference and taste

A series of 175 college-age subjects checked food dislikes on an alphabetical listing of 120 foods previously tested on a college-age Ohio population. Excluding foods unfamiliar or never sampled by each subject, the percentage of familiar foods disliked was used as the individual score. Scores ranged from 0 to 55 per cent; the median was 10 per cent, as in the standard population. The only significant correlation between per cent of food dislikes and taste thresholds was obtained with quinine, which indicates that the number of foods disliked grows in proportion as the taste threshold for quinine and other "Gaussian" compounds decreases (Fischer et al. 1963; Fischer and Griffin, 1961 b; Fischer et al. 1961 a).

Actually, for that 25 per cent of subjects on the right end of the Gaussian distribution in Fig. 6 with high taste thresholds (7-8 and above) for quinine, few or no food dislikes can be predicted reliably. Similarly, for 25 per cent of the subjects on the other end of the Gaussian distribution with low quinine thresholds (4 and below), it is easy to predict that a number of foods will be disliked. It is for these discriminating people that the "Great Gourmet" Brillat-Savarin, wrote his **Physiologie du goût** in 1825. For those subjects with quinine thresholds 4-7, who constitute the 50 per cent middle range of the Gaussian distribution curve, no prediction can be made as to number of foods disliked in relation to taste thresholds. In this group, cultural, social and idiosyncratic variables probably will decide the matter of food aversions (Fischer et al. 1961 a; Smith et al. 1955). Our 120-item food-chart, therefore, should not be indiscriminately applied, since such factors as age, ethnic composition, extent of exposure to a variety of foods, religion, socioeconomic status and other factors vary from population to population. For each population and geographical area, therefore, characteristic food charts will have to be designed.

(III) Smoking habits and taste

Interestingly, sensitive quinine taste responders, who are more particular in their choice of food, are also more"sensitive" toward smoke, that is, we find fewer heavy cigarette smokers (twenty or more cigarettes per day) among them than among insensitive quinine taste responders. These findings, first established on 127 college-age students (Fischer et al. 1963), were extended and confirmed with the 31-50-year age range (Kaplan et al. 1964). The same relationship was reported by Krut, Perrin, and Bronte-Stewart (1961), who used a different taste-testing methodology.

The question should be raised now about the relation of age, sex, smoking habits and taste responsiveness. The common misconception that thaste threshold and age increase together is probably based in part on Harris and Kalmus (1949), who found that "the perception of PTC (phenylthiocarbamide), like that of any other sensation, decreases with age. This finding has been repeatedly published (Kalmus and Hubbard, 1960), and appears to have become common "knowledge", in spite of the fact that Kalmus did not control for smoking. When we controlled for smoking, no significant age- or sex-related differences in taste sensitivity for quinine or 6-**n**-propylthiouracil were observed in a sample of 268 **non-smoking** subjects 16-55 years of age. Among 127 "heavy smokers" (twenty or more cigarettes per day), however, a significant deterioration in taste responsiveness was apparent with increasing age. Aging and smoking, either of which alone does not dull taste responsiveness, together exert a progressive reduction in taste responsiveness (Kaplan et al. 1965). It seems advisable, therefore, to control for age and sex when investigating factors that may be related to taste responsiveness in smokers. Since up to the twentieth year taste responsiveness is unaffected by smoking, control for smoking habits is required in studies involving groups older than twenty. The influence of age on taste thresholds for quinine and PROP, in male and female non-smokers and heavy smokers, is illustrated in Fig. 10, 11, 12 and 13.

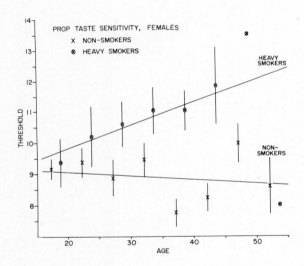

Fig. 10. Influence of age on taste threshold for PROP
(6-n-propylthiouracil) in female heavy smokers
and non-smokers. [As for subsequent fig. 11,
12, 13, the mean score and standard error for
each group are shown, together with the fitted
regression line for the scores of subjects aged
16 to 55 years. Thresholds are based on solution
numbers. Vertical lines represent ± 1 standard
error.]

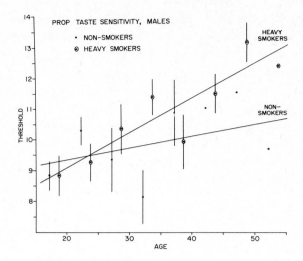

Fig. 11. Influence of age on taste threshold for PROP
(6-n-propylthiouracil) in male heavy smokers
and non-smokers.

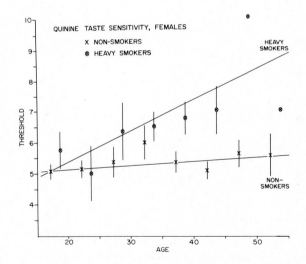

Fig. 12. Influence of age on taste threshold for quinine
in female heavy smokers and non-smokers.

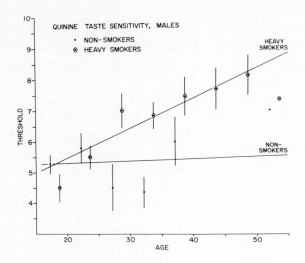

Fig. 13. Influence of age on taste threshold for quinine
in male heavy smokers and non-smokers.

We found taste insensitives as a group to be MBTI sensors (down-to-earth organizers) as well as extroverts and, therefore, heavy smokers. Eysenck also found that the majority of the thousands of smokers he has tested are extroverts (Eysenk, 1965). As expected, Kay reports that the oral contraceptive users among the 32,000 women he tested are more likely to be smokers and to smoke heavily than non-users (Kay, 1969). On the basis of the relation we established between smoking, high taste threshold and personality type, one would expect these women smokers to be good administrators, or sensors in terms of the MBTI : down-to-earth organizers of their reproductive cycles.

(IV) Taste and reactivity

We have described taste-sensitive subjects as sensitive in many other respects as well. Taste sensitivity, drug sensitivity, etc., are all manifestations of a general sensitivity, of which personality traits (Fischer et al. 1965) are only the most obvious (Corlis et al. 1967). For instance, extreme introverts secrete almost 1 g more saliva when stimulated by lemon juice than do extreme extroverts, who show little or no increase in salivation (Eysenck and Eysenck, 1967). Evidently, not only do extroverts have high taste thresholds, but also higher thresholds for other sensory modalities, i.e., the level of physical intensity required for them to perceive stimulation is higher than in introverts. In other words, they display low levels of arousal as evidenced by their high amplitude, low frequency wave forms on the electroencephalograph (EEG) (Gale and Coles, 1969). This is in line with the well-known gregariousness of **extroverts**, which can now be accounted for in terms of **low arousal** levels of a subject in need of stimulation because he cannot remain alert without it. Arousal can also be expressed as the ability to sustain vigilance without distraction, and extroverts are usually not as good at tasks requiring stability of attention. No wonder, then, that extroverts are more affected by hallucinogenic drugs, as measured by increased variability on their EEG and increased reaction time while under the drug. Tarriére and Hartemann (1964) found that, in a visual vigilance task which combined detection of peripheral signals with tracking, best performance was given by the introvert smokers, while introvert non-smokers performed at about the same level as extrovert smokers. Both the higher arousal level of introverts and the stimulating effect of nicotine * on the central nervous system (Goodman and Gilman, 1941) may account for the reported phenomena. The level of arousal, then, is apparently reflected in the quality of the vigilance performance, since performance decreased in this order : introvert smokers, introvert non-smokers and extrovert smokers, and extroverts deprived of smoking at the time of the test.

We also find that intuitive and introvert subjects (the majority of our sensitive tasters and drug reactors) are also frequent dreamers, dream more in color than in black-and-white, prefer color to form (Fischer et al. 1969), and display high mean energy content as well as greater coefficients of variation, an index of variability, on the EEG. In contrast, our down-to-earth sensor and extrovert subjects (the insensitive tasters and insensitive drug reactors) dream infrequently, dream rather in black-and-white than in color, prefer form to color (Fischer et al. 1969), and display low mean energy content as well as lower coefficients of variation on the EEG (Thatcher et al. 1970).

While we tend to emphasize the relation between MBTI intuition-introversion and chemoreceptive (taste) **sensitivity**, on the one hand, and sensation-extroversion and **insensitivity,** on the other, the leitmotif of Eysenck's research is the correlation of extroversion and introversion, in terms of the Maudsley Personality Inventory, with level of cortical arousal. Although Eysenck's extraversion-introversion types apparently do not vary with age or sex (Eysenck and Eysenck, 1970), and the MBTI intuition-sensation and extroversion-introversion continua are based on Jungian concepts which to some extent depend on the age-related maturation process, the two approaches to personality are basically similar, and we shall use the respective data interchangeably.

According to Eysenck, extroverted people smoke mor cigarettes, drink more coffee and alcohol, consume more sugar, eat more and spicier foods, have sexual intercourse more frequently and from an earlier age, and generally seek stronger sensory stimulation than do introverts (Eysenck, 1965). Extroverts also show greater pain tolerance, less stimulus deprivation tolerance and shorter perceptual after-effects (Eysenck, 1963). Correlated with this, though not very highly, is body build (Fischer et al. 1966) : extroverts are broader and rounder, introverts longer and leaner. Furthermore, there are undoubtedly complex differences in the central nervous system and cortex for individuals of these two types; introverts show greater cortical arousal, possibly mediated by reticular formation activity, extroverts show greater autonomic habituation and less vigilance (Eysenck and Eysenck, 1970).

We first uncovered the relation between taste and drug reactivity (Fischer and Griffin, 1963) because we looked at both as manifestations of the same organismic sensitivity on the chemoreceptive level, when it was only a short step to link this sensitivity with that of the personality (Fischer et al. 1965). Computerized factor analysis has recently allowed Eysenck to cement the link between introversion and higher cortical arousal and, most recently, with low sensory thresholds (Eysenck and Eysenck, 1970). His data confirm and elaborate his and our own earlier findings, and justify our view of perception-behavior as the symbolic interpretation of central nervous system activity (Fischer, 1968; Fischer and Warshay, 1968). Interpretation, therefore, cannot be separated from what is interpreted, and thus taste sensitivity, for instance, cannot be separated from a sensitive personality trait or high cortical arousal; all are manifestations of a particular type of systemic organization.

---

* After an initial excitatory phase, nicotine acts like a tranquilizer (Erlenmeyer, 1951).

(V) Variability and taste

I have mentioned how 50 per cent of a **population** on retest shows a drop in taste threshold (fig. 2). **Individual** variability, however, can be more exactly determined if semimicro taste threshold determinations are used. For instance, when we used the semimicro method on two trained, taste-sensitive volunteers, one always displayed a variability of about ± 0.4% when retested on any of the 28 compounds used, while the other always had a variability of ± 10.5% (Fischer, 1967 a).

Gustatory retest variability, however, is apparently only one of many expressions of organismic variability, for when we administered an oral dose of 160 µg/kg psilocybin to 15 middle-class, college-age volunteers (median age 23.5 years), their taste-threshold **retest variance** under no-drug conditions was significantly correlated with the psilocybin-induced (T_2-T_1) **behavioral change**, as measured by the Minnesota Multiphasic Personality Inventory (MMPI), which was administered before (T_1), at the peak of the subject's psilocybin experience (T_2). and again on the next (control) day. The behavioral change is brought about by the central, ergotropic-excitatory effects of this hallucinogenic drug. If the psilocybin-induced **behavioral change** on the MMPI and the magnitude of the subjects' taste-retest variance are rank-ordered, we obtain a correlation showing that the more stable a subject's taste threshold on retest, the more stable his behavior at the peak of the drug experience, i.e., the less psychopathology he shows as evidenced by low T_2-T_1 scores on the MMPI*(Fischer and Warshay, 1968; Fischer et al. 1968) (see Table 3).

Table 4 illustrates the significant correlations between perceptual and behavioral variables and their independence on 'dissociation' from the autonomic nervous system variable, drug-induced pupil diameter increase. We regard pupil diameter increase, which follows a drug dose-response relationship, as an indicator of ergotropic arousal – the activity level of that which is to be **interpreted** – and the perceptual-behavioral variables as manifestations of the perceptual-behavioral **interpretation** of that arousal. In our view, man is a self-referential system, interpreting his nervous system's activity in terms of perception-behavior. The 'dissociation' between the autonomic and perceptual-behavioral variables, therefore, illustrates the relative independence of interpretation from that which is interpreted (Fischer et al. 1970).

The most important implication of the data in Table 4, however, is that drug-induced **perceptual and behavioral change co-vary** (irrespective of the extent of drug-induced autonomic activity). The connecting hyphen between **perception-behavior** is based on these correlations, and implies the unity of the symbolic interpretive activity. Degree of perceptual stability, therefore, implies a similar degree of behavioral stability, and vice-versa (Fischer et al. 1969; Fischer, 1970; Fischer et al. 1970).

So far in this paper we have examined various correlates of gustatory **sensitivity**, relating taste threshold to other perceptual-behavioral threshold phenomena. We have also examined intraindividual perceptual and behavioral stability. Fig. 6 demonstrates the relation between sensitivity and degree of stability, showing that sensitive subjects (on the left tail of Fig. 6), based on their standard deviation (S.D.) on perceptual-behavioral tasks, can be subdivided into sta-

---

* We have confirmed this relation between perceptual stability at T_1 and behavioral stability at T_2 on a large variety of perceptual and behavioral tasks (Fischer et al. 1969);

Initials and sex of subjects N=15	Taste Retest Variance: Magnitude of change in molarity between first macro-quinine-taste threshold and semi-micro-retest	Rank order of taste retest variability	Rank order of drug-induced behavioral change measured with the MMPI	Σd^2 (T_2-T_1) K-corrected MMPI T-scores
W.R. ♂	121.88×10^{-6} M	15	15	2888
S.H. ♂	28.13×10^{-6} M	14	8	742
H.M. ♂	18.75×10^{-6} M	13	9	1006
(B.T. ♂)	12.89×10^{-6} M	12	3	297
R.M. ♀	9.38×10^{-6} M	11	13	2308
V.M. ♀	4.69×10^{-6} M	10	11	1052
B.N. ♀	3.22×10^{-6} M	9	12	1394
D.O. ♂	2.64×10^{-6} M	8	14	2521
A.M. ♂	2.34×10^{-6} M	7	5	372
S.J. ♂	1.17×10^{-6} M	6	4	339
S.S. ♀	1.17×10^{-6} M	5	7	502
S.A. ♀	0.58×10^{-6} M	4	6	444
K.T. ♂	0.58×10^{-6} M	3	2	280
H.S. ♂	0.58×10^{-6} M	2	1	77
(O.S. ♀)	0.00×10^{-6} M	1	10	1043

Table 3. Rank-order of perceptual (taste) threshold retest variability without drugs (T_1), correlated with rank-ordered psilocybin-induced (T_2) behavioral change ($T_2 - T_1$) as measured by the Minnesota Multiphasic Personality Inventory (MMPI). Perceptual data were obtained from forced-choice procedures. The Spearman rank-order correlation coefficient, corrected for tied ranks $r_s = 0.49$, and for a two-tailed test, $p < 0.5$. For details see Fischer et al. 1968.

The correlation is not impressive, but is more than suggestive in the light of other, similar data (Fischer and Warshay, 1968; Fischer et al. 1968). The reason that the correlation is not particularly high is that three months elapsed between the two taste threshold determinations. Evidence accumulated since these data shows that the magnitude of the variability on a perceptual test as T_1 is an **accurate and reliable** predictor of a subject's drug-induced perceptual-behavioral performance (at T_2), provided that both fall on the **same day**. In recent experiments, we have measured the standard deviation on handwriting area (SDHA) at T_1, usually at 10:00 A.M. (0 time), which is followed 90-110 minutes later, at drug peak, by a perceptual or behavioral test which takes from 5 to 60 minutes, but no longer. Under these conditions, for example, the size of the SDHA is (1) positively correlated with self-chosen tapping-rate/second, a five-minute test, at T_1 (N = 17, r = 0.604, p < 0.01) and (2) negatively correlated with the duration of the 63-minute time estimation task at T_2 (N = 11, r = -0.704, p < 0.01). We routinely use SDHA because it is quick and simple (Fischer, 1970).

N=15 SUBJECTS	DEGREE OF DRUG INDUCED INCREASE IN PUPIL DIAMETER	MAGNITUDE OF STANDARD DEVIATION OF SPATIAL DISTORTION THRESHOLD AT T_1		TASTE RETEST VARIANCE	MAGNITUDE OF GOLDBERG FORMULA SCORES AT T_2	MMPI DRUG-REACTIVITY SCORES
		PHOROMETER BASE DOWN	PHOROMETER BASE UP			
	COLUMN 1	COLUMN 2	COLUMN 3	COLUMN 4	COLUMN 5	COLUMN 6
WR ♂	6	8	11	15	14	15
DD ♂	6	15	14	8	15	14
RM ♀	11.5	3	2	11	8	13
BN ♀	9.5	12	10	9	13	12
JM ♀	6	10	6	10	11	11
DS ♀	14	9	9	–	3.5	10
HM ♂	2	14	7	13	10	9
SH ♂	9.5	11	15	14	12	8
SS ♀	14	6	4.5	5.5	2	7
SA ♀	11.5	7	13	3	6	6
AM ♂	6	4.5	8	7	3.5	5
SJ ♂	14.	4.5	4.5	5.5	5	4
BT ♂	6	13	12	12	9	3
KT ♂	2	–	3	3	7	2
HS ♂	2	2	–	3	–	–
	r_s = .68(p<.01)		.59(p<.05)	.72(p<.01)	.68(p<.01)	
	r_s = .00		.11	-.17	-.18	-.11

Table 4. Spearman correlation coefficients (r_s) illustrating the lack of relation between the autonomic variable pupil-diameter increase (column 1) and perceptual and behavioral variables (columns 2-6), and the presence of significant correlations between the perceptual and behavioral variables. Rank orders listed are : degree of drug-induced increase in pupil diameter (column 1), magnitude of the standard deviation of the spatial distortion threshold measurements made pre-drug with phorometer base down (column 2) and phorometer base up (column 3), magnitude of taste-retest variance under pre-drug conditions, magnitude of Goldberg formula score at drug-peak (column 5), and MMPI drug-reactivity (column 6).

ble and variable subjects. The insensitives, on the right tail, similarly fall into perceptual-behaviorally stable and variable subgroups.*

While **sensitivity** denotes higher activity baselines, **stability** (a small variance on taste-retest, a small S.D. on a yet unlimited variety of perceptual or behavioral tasks) relates to predictability of perceptual-behavioral performance.

GENETIC ASPECTS OF CHEMORECEPTION

Most of the pedigree and population studies on taste thresholds have been limited to the bitter PTC-type (PTC = phenylthiocarbamide, i.e. phenylthiourea) antithyroid compounds that contain the characteristic $H\text{-}N\text{-}C = S$ grouping. Taste thresholds for these approximately forty compounds (Harris and Kalmus, 1949 a; Barnicot et al. 1951; Harris, 1955) (such as PTC, 6-**n**-propylthiouracil [PROP], 1-methyl-2-mercaptoimidazole, etc.) follow a **bimodal distribution** in a population**, differentiating it into tasters and "non-tasters". The classical hypothesis of "non-tasting" is based on homozygosity for simple Mendelian recessive alleles (Harris, 1955).

It should be emphasized, however, that taste sensitivity to PTC-type compounds (Fox, 1932) is a special case, since taste thresholds for most compounds do not follow a bimodal but rather a Gaussian or monomodal distribution. The importance of compounds with a Gaussian distribution is emphasized by our observation that subjects with low taste thresholds for one such compound, such as quinine, also display low taste thresholds for all other such "Gaussian" compounds (except hydrochloric acid) (Fischer and Griffin, 1964 a; Fischer et al. 1965) irrespective of their taste quality (Fischer and Griffin, 1964 a; Fischer et al. 1962; Fischer and Griffin, 1963). However, a sensitive quinine taster, who is therefore also a sensitive taster of sucrose, urea, chlorpromazine, etc. **may or may not** be a "non-taster" of PTC-type compounds. Similarly a very insensitive quinine taster may or may not be a "non-taster" of PTC-type compounds (Fischer and Griffin, 1964 a; Fischer et al. 1962; Fischer and Griffin, 1963; Fischer and Griffin, 1964 b; Kaplan and Fischer, 1965). For these and other reasons (Fischer et al. 1966) we classify the mediators of gustatory chemoreception as **"Gaussian" sensors**, i.e., sensors of chemicals with unrelated structures, and **bimodal receptors**, i.e., receptors of compounds with a specific chemical structure characterized by the presence of the $H\text{-}N\text{-}C = S$ molecular grouping.

PTC has a strong aromatic odor which may account for the clean but apparently artefactual separation of the tasting and "non-tasting" modes in a population. Our observations of the influence of PTC odor on its taste thresholds have been confirmed by Skude (1963), who found that 65 per cent of his subjects displayed higher thresholds for PTC when their nostrils were plugged. In light of our own data, as well as that of the literature, the ability to taste odorless PTC-type compounds is, therefore, neither an all-or-none phenomenon nor a typical continuous variable.

---

* The terms 'stable' and 'variable' denote subjects with lower and higher cortical arousal, i.e., subjects in need of more sensory stimulation (stable extrovert-like 'maximizers') or less stimulation (variable introvert-like 'minimizers'), respectively (Fischer et al. 1969).

** This bimodal distribution is present in both mice and men, both of whose taste sensitivities for bimodally distributed compounds are determined largely by a single autosomal locus (Klein and DeFries, 1970).

We use exclusively an odorless PTC-type compound, 6-**n**-propylthiouracil (PROP) (Fischer and Griffin, 1964 a; Kaplan and Fischer, 1965; Trotter, 1962; Fischer and Griffin, 1962; Fischer et al. 1961 b), which is the least toxic of the phenylthioureas (Trotter, 1962) and the taste thresholds of which display a high positive correlation with those of PTC (Fischer et al. 1965a). Hence we can reliably test for taste while controlling for smell.

Two examples will illustrate the confusion created in the literature by unreliable methodologies and by the choice of an inappropriate compound when trying to link the genetically controlled trait of non-tasting of PTC-type compounds with certain somatic or behavioral traits.

(1) Chung et al. (1964) determined the tasting or non-tasting status of subjects with paper soaked in PTC solution. Although the authors reassure themselves that there is only a negligible discrepancy between their paper test and a solution technique (Hartmann, 1939), they do not mention that the "Hartmann solution technique" does not use R. Fisher's double-blind, forced-choice sorting procedure (Fisher, 1951), a **sine qua non** of reliable taste-testing (Fischer and Griffin, 1964 a; Fischer, 1967 a) which has been used by Harris and Kalmus and, in a more refined form (Fischer and Griffin, 1962) by ourselves.

(2) While the previous authors at least tried to back their preliminary results with the "Hartmann solution technique", two researchers from Johns Hopkins University based their genetic research on a single test with PTC paper and found a "strikingly high proportion of PTC tasters among heavy cigarette smokers" (Thomas and Cohen, 1960). We have demonstrated above that retesting is another **sine qua non** of obtaining a reliable taste threshold distribution. Of course, a retest with an already unreliable test method would have shown the authors that their own data are unreproducible (cf. Fischer et al. 1963)*

The relation between smoking and high taste thresholds for bitter-tasting "Gaussian" compounds such as quinine is a complex one to begin with. For instance, it is impossible to establish a significant correlation among college students – but only a trend – since this group constitutes a slanted sample, being composed of 60 per cent MBTI intuitors (MacKinnon, 1967), i.e., sensitive tasters. Another difficulty arises when one considers that, although high taste thresholds for quinine are correlated with smoking and low amounts of food dislikes (Fischer et al. 1963; Fischer and Griffin, 1961 b; Fischer et al. 1961 a), this relation does not hold for the "Gaussian" compounds sodium chloride, hydrochloric acid and sucrose (Fischer et al. 1963; Krut et al. 1961). The apparent contradiction is resolved when one considers that the range of **quinine** taste responsiveness for both sensitive and insensitive subjects is identical with the solubility range of the compound which for quinine spans twelve solution numbers representing a **4096-fold difference in concentration** (see fig. 6). However, for **other** "Gaussian" compounds such as sodium chloride, sucrose, hydrochloric acid and chlorpromazine (Fischer et al. 1962), this correspondence is lacking; there is a surprisingly narrow range of taste responsiveness for these much more water-soluble compounds, a taste responsiveness which is limited to ranges representing only **512, 512, 128 and 64-fold differences in concentration**, respectively (cf. Fischer et al. 1963). Evidently these narrow ranges do not permit such a clean differentiation between sensitive and insensitive taste responders as is possible with quinine.

---

* Even a retest with PTC solution instead of paper results in a low retest reliability, as shown by R. Guttman et al. (1965). While taste studies using PTC solution and a sorting-out technique yield reproducible taste thresholds, the antimode is artefactually clean due to the odor of PTC.

Jörgensen has compiled some 300 studies (up to 1967), listing the geographical distribution of the PTC non-tasting trait (Jörgensen, 1969). However, the validity of most of these studies is doubtful, since at least one of the following **sine qua non**'s of reliable taste-testing was observed : (1) the use of an odorless PTC-type compound such as 6-n-propylthiouracil (PROP); (2) the use of a solution sorting-out technique based on Fishers's design; and (3) validation through retest.

*   *   *

We have emphasized two criteria for classifying a subject in terms of taste sensitivity (see fig. 14). The first general criterion classifies him according to taste threshold for some 100,000 "Gaussian" compounds such as quinine, sucrose, sodium chloride, etc., while the second, special criterion classifies him according to the taster – "non-taster" trait for some 40 bitter-tasting PTC-type drugs, all of which display a more or less distinct antithyroid activity.

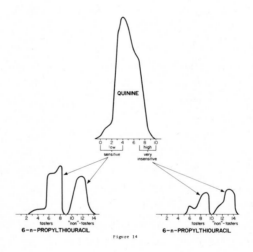

Fig. 14. Distribution of quinine macro taste thresholds in a population (N=228), above. Below: distribution of 6-n-propylthiouracil (PROP) taste thresholds in the same population (Fischer and Griffin, 1964 a). Note that the position of the PROP antimode is influenced by the subjects' quinine thresholds. The PROP antimode is at threshold (solution number) 9 for subjects with low quinine thresholds; while it is at 10 for insensitive subjects with high quinine thresholds. A glance at the figure also reveals the total lack of very sensitive PROP tasters – with thresholds 2-5 – among the very insensitive quinine tasters. The influence of quinine taste sensitivity on the expression of PROP responsiveness may be regarded as an example of partial epistasis in humans. Over 80 per cent of the subjects were retested within ten days.

It is apparent from fig. 14 that the degree of sensitivity for "Gaussian" compounds such as quinine influences the expression of taste sensitivity for PTC-type compounds such as PROP, although the latter is genetically controlled. Fig. 15 illustrates this by showing the significant **difference in mean quinine macro taste threshold** between PROP tasters and PROP "nontasters" (Fischer and Griffin, 1964 a). Fig. 16 shows that the mean PROP taste threshold for each of the tasting and non-tasting modes is significantly higher for the very insensitive tasters of quinine than for the sensitive tasters. Note that the **antimode** dividing the quinine-sensitive

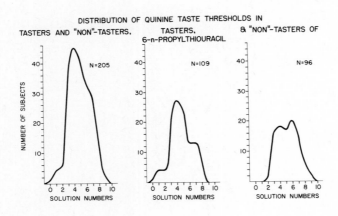

DISTRIBUTION OF QUININE TASTE THRESHOLDS IN

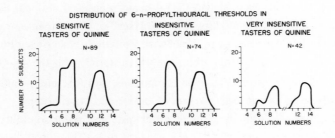

DISTRIBUTION OF 6-n-PROPYLTHIOURACIL THRESHOLDS IN

Fig. 15 and 16. The numbers on the abscissa refer to thresholds in solution numbers expressing doubling concentrations (N=203), (N=109), (N=96, in fig. 15; (N=89), (N=74), (N=42) in fig. 16. Because of the "shifting antimode" phenomenon, subjects with a PROP threshold at the antimode (No. 9 for sensitive and insensitive quinine tasters and No. 10 for very insensitive quinine tasters) are omitted.

15

PROP tasters and "non-tasters" **shifts** from **PROP** Solution No. 9 in the sensitive group to solution No. 10 for the very insensitive tasters. These relationships preclude the existence of a
static antimode for PROP, since the position of the antimode for PTC-type compounds is governed by the subjects' quinine sensitivity. For example, a subject with PROP threshold 10 is
a "non-taster" of PROP if his quinine threshold is low (0-4), whereas if his quinine threshold
is high (7 and above), he is a taster of PROP. Accordingly, the antimode dividing PROP tasters
from PROP "non-tasters" is at Solution No. 9 for sensitive quinine tasters and at 10 for the
quinine insensitives. Because of this "shifting antimode" phenomenon, subjects with a PROP
threshold at the antimode were omitted from the presentation of data in Fig. 15 and 16. It is
regrettable that the "shifting antimode" has so far been ignored in studies of the genetic mechanism of the PTC non-tasting trait.

PTC and PTC-type compounds are drugs, i.e., water-soluble, bitter-tasting 'colorless
dyes', with a characteristic $-N-C=S$ grouping and a characteristic antithyroid activity. Recently, Goedde and Ohligmacher (1965) studied the taste threshold distribution of another PTC-
type compound, anetholtrithione (ATTH):

ATTH contains the $-S-C=S$ (or alternately $=CH-C=S$) instead of the $-N-C=S$ grouping, with a
reported trimodal taste threshold distribution in which two of the modes were approximately
those of the classical bimodal distribution for PTC and PROP (Goedde and Ohligmacher, 1965).
Dawson et al. (1967), however, claim that ATTH displays a classical bimodal taste threshold distribution, and that the reported trimodality is an "artifact" of the 30 per cent ethanol solvent
which "dulls the taste" but is necessary to dissolve this relatively water-insoluble compound.
The $=CH-C=S$ and the $-S-C=S$ groupings are, however, isosteric* with the $-N-C=S$ grouping
which characterizes the bitter-tasting PTC-type compounds, and that one would therefore expect them to be both drugs and genetic markers of tasters and "non-tasters".

Our definition of a drug in terms of bitter taste and water solubility includes of
course the PTC-type genetic markers. One of the classical rules-of-thumb in designing drugs
with a particular pharmacological activity is isosteric substitution. Although one cannot predict whether a given substitution will enhance or diminish a compound's pharmacological activity, the fact that ATTH is not a particularly suitable genetic marker of the non-tasting trait

---

* Atoms or ions having identical numbers of electrons in their outermost shell are termed isosteric (Erlenmeyer, 1951).

for PTC-type compounds implies that it is not a particularly active antithyroid drug, either. Moreover, ATTH does not fit our criterion of water solubility for drugs, and therefore justifies the definition itself. In short, based only on the solubility characteristics of ATTH, our definition of a drug predicts the insufficient activity of ATTH as an antithyroid drug and hence as a clean genetic marker.

We have recently presented and reviewed some genetic aspects of chemoreception in man (Fischer, 1967 a; Kaplan et al. 1967), from which the distribution of average intrapair taste threshold differences for hydrochloric acid, quinine and PROP in the monozygotic and same-sex dizygotic twin pairs illustrated in fig. 17. There is no difference for **hydrochloric acid** * between pairs of genetically identical twins and pairs of fraternal twins, a small but significant difference for **quinine,** while **PROP** displays the largest intrapair difference. Therefore, genetic factors play a predominant role in determining the taste sensitivity of individuals for PTC-type compounds such as PROP, a small role in determining sensitivity for quinine, and no role at all in HCl taste. If 1.00 in terms of the Holzinger index represents 100 per cent heritability, the Holzinger index (H) values 0.33 for quinine and 0.87 for PROP are meant to express the extent of genetic control on the taste thresholds for the two compounds.** The low H value for quinine implies that taste sensitivity for the hundreds of thousands of "Gaussian" compounds is under little or no genetic control, yet, as we have seen, these 'environmentally determined' sensitivities **apparently** exert a strong influence on the genetic expression of taste sensitivity for PTC-type

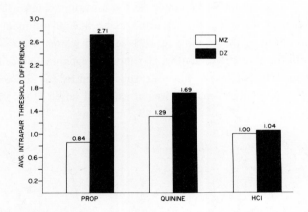

Fig. 17.   Average intrapair threshold differences in monozygotic (MZ) and same-sex dizygotic (DZ) twin pairs for HCl, quinine, and PROP. Taste thresholds were determined in 45 dizygotic (DZ) and 26 monozygotic (MZ) twin pairs for HCl, and in 70 DZ and 75 MZ twin pairs for quinine and PROP (in that order). The latter 145 twin pairs were supplemented with 191 combinations of non-twin sibling (SIB) pairs of which 142 were also taste-tested for HCl. For the sex distribution of twins and siblings, methodological details, and a complete two-factor (kinship and sex) analysis of variance of the data, the reader is referred to the original study (Kaplan et al. 1967).

---

*   Hydrochloric acid is the only "Gaussian" compound so far uncovered the thresholds of which are not correlated with those of other "Gaussian" compounds.

** For a discussion of the suitability of the Holzinger index as a measure of heritability see Fischer (1967 a).

compounds, as shown by the "shifting antimode" phenomenon. But it is arbitrary to state that taste sensitivity for "Gaussian" compounds exerts an influence on the genetically-determined taste sensitivity for bimodally distributed compounds, since taste sensitivity for "Gaussian" compounds, drug reactivity, personality traits, etc., are **all** manifestations of a general **organismic** sensitivity, and thus introversion-intuition or higher cortical arousal are as important "influences" on the "shifting antimode" as are low taste thresholds for quinine.

How otherwise could we account for our observations that 17 out of 20 married couples we tested between 1962-1964 had similar quinine taste thresholds, and that nearly all of about 40 pairs of good friends displayed the same range of taste sensitivity? Because we were at a loss to account for these data we did not publish them, but now the relations between taste sensitivity, drug sensitivity, certain personality traits, cortical arousal, reaction time, food dislikes, smoking habits, etc., suggest that apparently sensitive people prefer sensitive people, and insensitive people prefer insensitives. This kind of 'natural selection' commonly ruins 'representative samples', and may in part account for our data on couples and friends, as well as for the predominance of certain personality types (and hence taste thresholds) within certain vocational categories.

It is conceivable that practice-induced **learning** (Fischer and Griffin, 1964 a; Fischer, 1967 a) on the taste sensitivity level and observational learning on the personality level may also play a role in these phenomena. Since we know that practice-induced learning occurs when semi-micro or micro taste-testing methods are used, the sharing of a sensitive taster's living habits by a less sensitive taster might result in a lowering of taste thresholds and a change of personality.

Pregnancy is a condition which results in the lowering of taste thresholds without practice (Fischer and Griffin, 1964 a), whereas stress, as we have mentioned, raises taste thresholds. Fig. 18 illustrates both of these points in a 21 year-old woman whose quinine thresholds were measured biweekly for about five months. Her initial pregnancy-induced increase in taste sensitivity (decreasing taste thresholds) is, however, superceded by the stress of a polyhydramnios pregnancy which resulted not only in marked weight gain and raised blood pressure, but also in

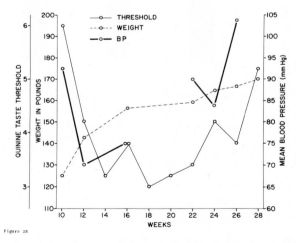

Fig. 18   The influence of stress (a polyhydramnios pregnancy) on quinine taste sensitivity determined biweekly by using the semimicro method. Note that the initial pregnancy-induced decrease in taste threshold is eventually raised above the initial level. Accompanying data include occasional weight measurements in pounds and mean systolic/diastolic blood pressure in mm Hg.

a decrease in quinine taste sensitivity. The pregnancy-induced increase in taste sensitivity was also accompanied by a decreased salivary oxidation rate, an expected result since taste sensitivity is correlated with low salivary oxidation rates (Fischer and Griffin, 1964 a; Fischer et al. 1963). Such a decrease in sensitivity can also be provoked by a stressful examination and loss of sleep, as we have repeatedly observed in college students.

VIVE LA (JUST-NOTICEABLE) DIFFERENCE !

Lowest taste thresholds in man for **all** compounds * are temperature-dependent with optimum thresholds obtained at 22^0 (Griffin, 1966; Fischer and Kaelbling, 1967). A characteristic feature of this temperature-threshold relation is a V-shaped, symmetrical curve (Fig. 19 displays such curves for quinine and PROP) with a sharp temperature-inversion at 22^0C.** While taste thresholds are a function of the chemical structure of the compound tested, the size of a just-noticeable taste difference, or jnd, is independent of a compound's chemical structure, its taste quality, solution temperature (within the range $17-27^0$C) and baseline solution concentration. These and other considerations indicate that taste thresholds are related to local chemoreceptor phenomena (such as affinity), whereas the size of a jnd reflects the state of excitation or tranquilization of the organism. In other words, a taste threshold reflects permanent features of organismic **sensitivity** (such as taste sensitivity, drug reactivity, personality type, etc.) while the jnd reflects the organism's level of **arousal** at the time of testing.

Indeed, we have found in healthy subjects that, both at the peak of psilocybin-induced ergotropic arousal and hyperthermia as well as in 'naturally' aroused acute schizophrenic patients, the jnd decreases – as measured by our micro method and expressed as $\Delta S/S$ or Weber ratio – whereas the jnd increases under phenothiazine tranquilizer-induced hypothermia (Fischer and Kaelbling, 1967; Fischer et al. 1965 b; Fischer et al. 1969). That the jnd actually indicates a system's level of arousal, and is hence independent of periodic fluctuations, has recently been confirmed while studying the temporal stability of the jnd over several months in six young and apparently healthy subjects (Fischer et al. 1970).

Schönpflug's finding that reticular activation also intensifies perceptual quality in modalities other than taste supports our finding that the size of the Weber fraction $\Delta S/S$ (signal to noise ratio) is a function of the system's level of arousal. This is not surprising in view of Kimura's (1961) and Chernetski's (1964 a) data that systemic sympathetic stimulation enhances afferent activity as recorded from the glossopharyngeal nerve during gustatory stimulation. Chernetski (1964 b) found that tactually elicited leg flexion in the frog is facilitated by the sound of a bell preceding the effective stimulus by 0.1-0.2 seconds. While the bell alone elicits no overt response, sympathectomy markedly reduces the intersensory facilitation of the leg-flexion reflex. Chernetski postulates a vertebrate arousal process which depends upon sympathetic outflow and which may be opposed by a central inhibitory mechanism.

---

* Semimicro thresholds for dozens of compounds representing a variety of chemically unrelated classes have been tested from 7^0 to 42^0in 5^0 intervals.
** Sato's claim – illustrated by Shimizu, Yanase and Higasihira's curves without reference, data or method – that most subjects show 'maximal sensitivity' to a variety of compounds between $30-40^0$C is unacceptable (Sato, 1963).

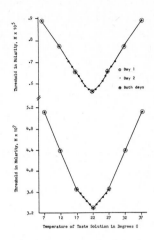

Fig. 19. Threshold-temperature dependence for 6-n-propylthiouracil
(top) and quinine sulfate (bottom) in a college-age subject.
These examples illustrate a V-shaped threshold-temperature
dependence with an optimum at 22°C characteristic for all
compounds, differing only in the slope of the curves. (Data
from our laboratory and dissertation of F. Griffin, 1966).

These data account not only for the wide variability once ascribed to gustatory
Weber ratios (Schutz and Pilgrim, 1957), but also for our **inability to duplicate** (Fischer et al.
1965 b) **Lemberger's** 40 successive **jnd** values for sodium saccharinate (Lemberger, 1908).
Lemberger's data have been repeatedly used to support models and mechanisms in gustatory
chemoreception. For example, while Lemberger's (1908) taste-response plot closely follows a
logarithmic curve deviating from the Fechner law only in the lower and higher ranges, Beidler
(1958) used Lemberger's data to fit his fundamental taste equation, whereas Duncan (1963)
used them for the demonstration of a "hyperbolic, rather than a logarithmic, relationship between the magnitude of the response and the stimulus intensity". Chernetski (1965) suggests
that our inability to replicate Lemberger's jnd data may be due to the slow procedure inherent
in the Lemberger technique, and the fact that her state of sympathetic arousal,and thus her
jnd size, might have varied over the many days of the experiment, whereas our semimicro and
micro procedures involve a much shorter time period. Chernetski also thinks that the initial
"half-moon" in Lemberger's successive jnd series (Fischer et al. 1965 b) may indicate that peripheral sympathetic effects are exerted primarily during these responses to particularly low stimulus concentrations. We conclude, therefore, that since the level of arousal influences the size
of the Weber fraction, jnd values should be determined so as not to be affected by variations in
level of arousal. Another alternative would be to take into account the state of the system by
inserting **t** (body temperature) as an index of the system's level of arousal into the formula describing the relation between stimulus magnitude and sensation (Fischer and Rockey, 1967 and
1968).

SPECIFIC AND NONSPECIFIC ASPECTS OF CHEMORECEPTION AND THE INTERACTION OF OLFACTION AND GUSTATION

Our finding that unilateral sectioning of the chorda tympani is not followed by a significant change in either quinine or PROP taste thresholds (Fischer et al. 1966) has cast doubt on the specificity of localized innervation on the tongue, as well as on the need for intact glossopharyngeal or chorda tympani nerves, and large numbers of taste buds and receptors, for taste sensation. A year later, Robbins (1967) reported that the "electric taste" threshold is not much higher shortly after division of the chorda tympani than in the presence of intact gustatory fibers. These findings are provocative in the light of another study in which the vallate papilla in the rat was excised in order to determine whether the remaining lingual tissue could regenerate a papilla and taste buds (Zalewski, 1970). Taste buds* did appear at all postoperative time intervals, but only in the innervated tongues, demonstrating that epithelium other than that covering the vallate papilla can give rise to taste buds, and that taste-bud development does not depend on the presence of a papilla. The nerve-epithelial interaction that results in taste-bud formation, therefore, may not require the presence of the original gustatory epithelium or papilla. When taste buds are cross-innervated (e.g., when the IXth nerve is forced to innervate fungiform papillae on the anterior part of the tongue), the cross-regenerated nerve is changed in its relative responsiveness to different taste stimuli. Even under these conditions, taste preference-aversion behavior (for sucrose, sodium chloride and quinine sulfate) is apparently unchanged eight months later, and Oakley suggests some dominant non-lingual sources of taste input (Oakley, 1969), determined by inherent properties of the tissue from which the cells are formed : the epithelium at the front may differ from that at the back of the tongue (Oakley, 1967).

Olfactory perception, like vision and stereognosis, may be confined to a single hemisphere when the input is restricted to a single nostril (Gordon and Sperry, 1969). Odors can then be recognized only in the hemisphere on the same side as the stimulated nostril, as evidenced by a subject's ability to name odors from the left nostril (i.e., to the left, speech-dominant hemisphere), but not from the right nostril. That perceptual recognition of odors through the right nostril is present, however, can be ascertained by other appropriate tests.

Some of the questions raised by the preceding studies of gustatory innervation remind us that individual chorda tympani units in the rat confirm the existence of multiple sensitivity (Pfaffmann,1962). Hence, the sensitivity patterns seen in chorda tympani units do not depend only on branched afferent terminations on several taste cells. The sensitivity range of many individual receptor cells also appears to be multiple, which has also been suggested for cutaneous sensitivity, since many A and C afferent fibers are sensitive to both mechanical and thermal stimulation. Thus, the neural code for gustatory quality appears to utilize the neural frequency pattern across many neural elements. This concept was developed by Erickson et al. (1965) as an "across-fiber patterning" model, and implies that gustatory quality is encoded in the form of across-fiber activity patterns. Pfaffmann has recently pondered the possibility that sensitivities to the four taste qualities – sweet, sour, bitter and salty – are independent and randomly distributed among taste fibers. Indeed, the assumption that independent sensitivities to the four taste qualities are

---

* For reviews of the structure of vertebrate taste buds and the innervation of fungiform papillae, see Graziadei, 1969 and Beidler, 1969.

randomly distributed among innervating fibers allows Pfaffmann to describe distributions of sensitivities in both chorda tympani and glossopharyngeal nerve fibers in the rat, and thus to postulate the characteristic nonspecificity of mammalian taste fibers (Pfaffmann, 1962). In a summary of evidence for his idea that the relative, rather than the absolute, amount of activity in any one set of afferent fibers may determine the quality of sensation, Pfaffmann (1969) examines the implications of such multi-quality response at several levels : (1) on the purely peripheral level, i.e., from receptors to afferent nerve fibers; (2) from afferent to central nervous system; and (3) from afferent nerve-discharge pattern to sensation. Pfaffmann's emphasis on relative firing rates is a device by which dominance of the primary stimulus for any particular fiber is preserved, even though other stimuli may become effective as concentration is raised. Pfaffmann calls this pattern theory a "ratio theory", and believes that there is little difference operationally between it and the specificity view of von Békésy (1966).

Von Békésy (1964), using electrical stimulation through an electrode small enough to rest on a single fungiform papilla at the tip of the tongue, reported that the only tastes identified were the classical four, and that only one taste was reported from each papilla, **except** for some large papillae near the edge of the tongue. From these, two tastes, invariably salty and sour, were reported. Moreover, his absolute taste threshold curves for **salty and sour** and for the different papillae types (as a function of stimulus frequency) closely resemble each other, with a broad range of frequency response. The respective curves for **bitter and sweet** again resemble each other, but have different and narrower frequency ranges *. v. Békésy also reports a relation between taste quality and certain anatomical features of single fungiform papillae. In other experiments (von Békésy, 1966), he stimulated single papillae with droplets of solutions representing the four basic taste qualities, and found again, as with electrical stimulation, that most papillae were sensitive to a specific quality. Again as before, however, some large papillae near the edge of the tongue were sensitive to two solutions, and a few to more.

Dzendolet (1969), in his comparative evaluation of Pfaffmann's nonspecific patterning theory and v. Békésy's specific, "neoMüllerian" theory, finds not much disagreement between these two only apparently contradictory theories. He contends that their proponents deal with different experimental conditions, and finds their results valid and complementary. Dzendolet compares their differences with those of the "telephone theory" and the "place theory" of the auditory pitch mechanism. In humans the "telephone mechanism" is active only below 50 Hz. Between 50 and about 3000 Hz, both mechanisms appear to play a role, whereas above 3000 Hz, only the place mechanism signals pitch (von Békésy, 1963). Similarly, separate mechanisms may be operative in gustation for salty and sour as compared to bitter and sweet.

Note that electrically induced taste is different from chemical stimulation. For instance, when electrical sweetness produced by a gold point electrode is compared with the sweetness of a sugar solution, doubling of the electrode voltage requires an eightfold concentration increase in the sugar solution in order to match the two sensation magnitudes (von Békésy, 1964).

---

* Bitter and sweet are the only taste qualities for which the phenomenon of aftertaste (analogous to visual afterimage) has been observed. For example, we have observed repeatedly that an insensitive taster of PROP (and quinine) will experience a sweet aftertaste for hours after tasting high concentrations of PROP for only a few minutes. Bitter and sweet are also the first taste qualities to disappear in response to most anesthetics, followed by sour and sweet (in this order) (v. Skramlik, 1963), supporting v. Békésy's grouping of the four taste qualities.

*   *   *

Because of the interconnection of the anatomical structures of gustation and olfaction in man, some interaction between the two senses is inevitable. Indeed, there is a dramatic loss in the ability of subjects to identify salt and sugar solutions when they are rendered functionally hyposmic by stopping the movement of odorous molecules from mouth to nose through the nasopharynx with a stream of air flowing in the opposite direction, that is, certain flavors cannot be identified without nasal chemoreception (Mozell et al. 1969). The need for such cues in interpreting stimulus configurations is not restricted to chemoreception. A similar situation is found on the organismic level during the ingestion of a novel food or drug or during exposure to any novel stimulation which alters central nervous system activity and thus calls for a characteristic perceptual-behavioral interpretation (Fischer, 1969 b; Thatcher et al. 1970). Any such first experience is responded to in a manner which is atypical when compared to later responses (Fischer, 1969 c).

Just as there are two ranges of taste sensitivity to "Gaussian" compounds irrespective of chemical structure and taste quality (Fischer and Griffin, 1964 a, and 1963), so there appear to be two ranges of smell sensitivity which are independent of the chemical structure and odor of the compound tested (Brown et al. 1968). Brown, MacLean and Robinette report that, irrespective of sex, there is one "general factor" (contributing about one-third of the total variance and twice as large as a second factor) which is positively correlated with olfactory sensitivity to all compounds tested * (Brown et al. 1968). The second factor might be accounted for in terms of Amoore's stereochemical (Amoore, 1962) and Wright's vibrational theories (Wright, 1966 a, and 1966 b), both of which are based on specific chemoreception.** Although it seems fair to assume that most sensitive tasters are sensitive smellers as well, a high intraindividual correlation between the two senses is perhaps unnecessary since odor can apparently be a cue for taste.

*   *   *

---

*   The odorants used (Brown et al. 1968) were six examples of the primary odor classes proposed by Amoore (1962), plus ammonia, potassium cyanide and water, and were tested in distilled water solutions by a serial dilution method resembling that of Fisher (1951) . The important results would be even more acceptable if each odorant were **absorbed** for testing, as on filter paper, since Teranishi (1970) has shown that thresholds may vary by many order of magnitude for the same odorant if tested in aqueous solution. This may be so because a compound in solution represents a more complex system than in the absorbed (solid) state; there are some thirty configurations of water (Horne, 1970; Anon, 1970).

**   The earliest proponent of an atomistic theory of olfaction is Lucretius : "You may readily infer that such substances as agreeably titillate the senses are composed of smooth round atoms. Those that seem bitter and harsh are more tightly compacted of hooked particles and accordingly tear their way into our senses and rend our bodies by their inroads". (Lucretius, 1954).

Having directed most of our attention to various specific aspects of chemoreception, let us put it back into its organismic context.

" Mouth, throat and nasal cavity are so innervated, and so disposed in
' relation to one another, that **any or all of three sensory systems ope-
rate simultaneously** [italics added] in response to the same stimulus.
The cutaneous sensibilities of the mouth region react to texture and
astringency, as well as biting, burning, cooling and tingle. The sense
of taste is also obviously involved, but it is olfaction that furnishes
the most elaborate experiences connected with food flavor. Behavio-
rally, then, it is difficult to partition the chemical senses and even to
separate them from other oral sensations when the stimuli are placed
in the mouth" (Stone and Pangborn, 1968).

Evidently, no sense is an island entire of itself. Certain taste stimuli such as ^{14}C-labelled glucose can even quickly **diffuse** from the oropharyngeal cavity directly to the brain and could thus contribute chemical information additional to that mediated by taste receptors (Kare, 1969). The variability observed in this phenomenon (Kare, 1969) may well be related to the intraindividual reactivity * we have already reported in this paper.

On the other hand, von Buddenbrock (1958) argues for specificity :

" In two highly developed forms of animal life, the vertebrates and the
insects, the sense of taste and the sense of smell can be completely di-
vorced from each other and regarded as two distinct, mutually independent senses. In all vertebrates the olfactory nerve runs in the most frontal position from the nose to the central nervous system, while the fifth,
seventh, and ninth cranial nerves conduct gustatory stimulation from the
tongue to the brain. This clear evidence of the **separation of these senses** [italics added] is corroborated by physiological experiment. If we re-
move the entire forebrain of a fish it can no longer smell, but it can still
be conditioned to respond to taste stimuli".

Most recently, Thompson et al. (1970) claim that there are "number-counting" neuronal firing responses which are independent of stimulus modality. They report in the associative cortex a receptive field whose cells recognize numbers and count, like Galton, who used his chemoreceptors for counting :

" I taught myself to associate two whiffs of peppermint with one whiff
of camphor, three of peppermint with one of carbolic acid, and so on.
Next I practiced at some small sums in addition; at first with the
scents themselves, and afterwards altogether with the imagination of
them. There was not the slightest difficulty in banishing all visual and

---

*Hecht and Treptow (1961) have shown that even rats have different but individually characteristic reactivities (to a variety of stimuli and drugs) (Hecht et al. 1960), while Mitoma (1970) has recently provided evidence that an individual who metabolizes one drug slowly also metabolizes other drugs slowly. There is a 9-10 per cent difference in metabolic rate at all adult levels between men and women, the latter of whom burn the fire of their lives at a lower rate, and thus consistently outlive men (Fischer, 1966).

auditory images from the mind, leaving nothing in the consciousness besides real or imaginary scents. In this way, without, it is true, becoming very apt at the process, I convinced myself of the possibility of doing sums in simple addition with considerable speed and accuracy solely by means of imaginary scents" (Galton, 1968).

*   *   *

When a man's eye is photographed while he tastes various solutions, pupillary dilation, which is mediated by the sympathetic division of the autonomic nervous system, can be monitored to indicate taste differences too small for him to articulate (Hess and Polt, 1966). Pupillary dilation is related to the affective arousal value of the stimulus, and there is a close relationship between it and activity in the parietal cortex, an area related to taste stimulation. Hess has also found a relationship between pupillary activity and the visual cortex when pictorial material of varying interest is presented to subjects (Hess, 1965). It seems that not-yet-interpreted stimulus configurations – commonly called "unsolved material" (Bradshaw, 1967) – maintain the high plateau of dilation indicative of arousal. That the size of a jnd is a reflection of the intensity of systemic, ergotropic arousal implies that the jnd is independent of the chemical structure of the tasted compound and its solution temperature. Beidler's "fundamental taste equation" (Beidler, 1958), therefore, is inapplicable to jnd data. The numerical differences in thermodynamic values in this equation are actually generated by differences in molar threshold concentrations.

*   *   *

Man has appreciated the sexual stimulants musk and civet for a long time, and has therefore used them widely in perfumes. There is a structural similarity between civettone and the steroid ketone androstanone; the steroid hormone Δ^{16}– androsten– $3a$– ol even displays a musky smell. In view of the universal role of perfumes and pheromones for sexual information transfer (cf. Michael and Keverne, 1970; Murphy and Schneider, 1970), some degree of sex-dependent differential response to these compounds might be expected. Indeed, in a recent pilot study we found that a 0.75 per cent solution of "natural musk" in alcohol * absorbed onto strips of filter paper can differentiate without overlap between males and females. All males tested "liked" the odor "somewhat" on a graded rating scale, while females of the same age "disliked" the odor "somewhat", and a few subjects could not smell the odor in this concentration. Interestingly, there was no **strong** preference or aversion expressed by any of the subjects at this particular concentration, and concentrations higher than 2 per cent did not clearly differentiate the two groups.

How might such a sex-linked preference-aversion be accounted for? Cholesterol is oxidized in the body to 2-methylheptan-6-one, a ketone structurally similar to the musk compounds. Due to sex-specific differences in enzyme activity (Langecker, 1969), one would expect a sex-dependent rate of degradation for cholesterol and thus different systemic baseline concentration of the ketone. This, then, could account for the differential preference response to "natural musk" and related compounds as well. As to the pharmacological mediation of the arousal-response to musky compounds, Kimura et al. (1968) found increased catecholamine response to musk, and assume a β-adrenergic mechanism.

---

* Courtesy of Firmenich and Co., Geneva.

Much has been written about the role of the limbic system in olfactory, emotional and sexual behavior, so it is important to realize that the size of the limbic structures in primates is completely independent of that of the olfactory bulb, which undergoes a distinct reduction in the ascending primate scale. "If there exists an olfactory influence on the limbic system it is probably very low in the higher mammals" (Stephan, 1966). Girgis suggests that all the recent experimental anatomical and physiological data point to the conclusion that the amygdala is more naturally classed with the "limbic system" than with the "olfactory system" (Girgis, 1969). MacLean nevertheless emphasizes the proximity of the amygdala and the septum, which

" are involved in oral and sexual functions, respectively. The amygdala has strong connections with the septum and, as Nauta (1960) has shown, with the medial dorsal nucleus as well. In the continuation of our investigations it has been of interest to find that by stimulating parts of the amygdala one may obtain chewing and salivation, with partial erection occurring as a recruited response after many seconds of stimulation or as a rebound phenomenon after stimulation is terminated (Reis et al. unpublished). In other words, excitation in a region involved in oral mechanisms readily spills over into others concerned with genital functions.

If there is any neural offender that can be blamed for this situation, then the olfactory sense, more than any other, must be considered the culprit (MacLean, 1959). When an animal is viewed in the ordinary elongated position, the oral and anogenital regions are at opposite poles. A corresponding relation is maintained in the topographical representation of the body in the post-central gyrus of the neocortex. In the organization of the lower mammalian brain, however, Nature apparently found it necessary to bend the limbic lobe upon itself in order to afford the olfactory sense close participation in both oral and anogenital functions (MacLean, 1962).

Olfactory-gustatory sensitivity thus transcends the naso-oral region in its systemic and symbolic significance. When we therefore say in praise, "He has taste", we imply a style rooted in sensitivity and intuition which patterns one's dialogue with the outer world, a style whose very personal character was long ago recognized in the Latin **De gustibus non est disputandum!** and the French **Chacun son goût!**

ACKNOWLEDGMENTS

These studies were supported by grants M-4694 and M-2731 from the National Institutes of Health, U.S. Public Health Service, and a grant from the (Ford) Foundations' Fund for Research in Psychiatry, number 66-341, as well as NIH General Research Support Funds distributed by the College of Medicine of the Ohio State University.

We would like to acknowledge the following journals and/or publishers in connection with the reproduction of certain figures and a table : Fig. 8, Grune and Stratton, New York; Fig. 9, **Archives Internationales de Pharmacodynamie et de Thérapie**, Gand (Belgium); Fig. 10-13, **Journal of Gerontology**; Fig. 14, 17, The Johns Hopkins Press, Baltimore; Fig. 15-16, **Arzneimittelforschung,** Editio Cantor, Aulendorf in Würtenberg, Germany; Table 4, **Perspectives in Biology and Medicine**, University of Chicago Press.

My warmest thanks are due to all my associates, co-authors and assistants who, during the last eleven years, have made possible our multidisciplinary approach to taste. I am also indebted to the Medical School and the Department of Psychiatry of the Ohio State University, and to Ian Gregory, M.D., Chairman, for his continued support.

During the writing of this paper, I had the good fortune to be able to rely on the creative editorial assistance of Jim Scheib. Jim Kreutzfeld's (Medical Illustration, The Ohio State University) and Vernon Cady's (Photography, Department of Psychiatry, The Ohio State University) skillful craftsmanship is also gratefully acknowledged, together with the patient, cheerful and reliable assistance of Pam Furney in the preparation of this manuscript.

REFERENCES

Alexander, P. and Hudson, R. 1954. Wool : Its Chemistry and Physics, 362, Chapman and Hall, London.

Amoore, J. 1962. The stereochemical theory of olfaction, *Proc. Sci. Sec. Toilet Goods Assoc. 37,* Suppl. 1, 1-23.

Anderson, C. 1955. The effect of subliminal salt solutions on taste thresholds. *J. Comp. Physiol. Psychol. 48,* 164-166.

Anon. 1970. After anomalous water and polywater-cyclimetric water. *New Sci. 45,* 694.

Barnicot, N., Harris, H. and Kalmus, H. 1951. *Ann. Eugen. Lond. 16,* 119-128.

Bartoshuk, L., McBurney, D. and Pfaffman, C. 1964. Taste of sodium chloride solutions after adaptation to sodium chloride; implications for the "water taste". *Science 143,* 967-968.

Beidler, L. 1958. The physiological basis of taste psychophysics, Proc. 2nd Symp. Physiol. Psychol. 1-10, Office of Naval Research, Dept. of the U.S. Navy, O.N.R. Symposium Report ACR-30.

Beidler, L. 1969. Innervation of rat fungiform papilla, in : C. Pfaffmann, ed., Olfaction and Taste, Vol. 3, 352-369, Rockefeller Univ. Press, New York.

von Békésy, G. 1963. Hearing theories and complex sounds. *J. Acoust. Soc. Amer. 35,* 588-601.

von Békésy, G. 1964. Sweetness produced electrically on the tongue and its relation to taste theories. *J. Appl. Physiol. 19,* 1105-1113.

von Békésy, G. 1966. Taste theories and the chemical stimulation of single papillae, *J. Appl. Physiol. 21,* 1-9.

Bendit, E. 1966. Infrared absorption spectrum of keratin : Spectra of α-, β-, and supercontracted keratin, *Biopolymers 4,* 539-559.

Bradley, W. and Easty, J. 1951. The selective absorption of optical antipodes by proteins, 2. *J. Chem. Soc.* 499-504.

Bradley, W. and Easty, G. 1953. The selective absorption of optical antipodes by proteins, 2. *J. Chem. Soc.* 1519-1524.

Bradshaw, J. 1967. Pupil size as a measure of arousal during information processing. *Nature 216,* 515-516.

Brodie, B. and Reid, W. 1969. Is man a unique animal in response to drugs? *Am. J. Pharmacy 141,* 21-27.

Brown, K., MacLean, C. and Robinette, R. 1968. Sensitivity to chemical odors. *Human Biol. 40,* 456-472.

von Buddenbrock, W. 1958. The Senses, p. 108, Univ. Mich. Press, Ann Arbor.

Chernetski, K. 1964 a. Sympathetic enhancement of peripheral sensory input in the frog. *J. Neurophysiol. 27,* 493-515.

Chernetski, K. 1964 b. Facilitation of a somatic reflex by sound in *Rana clamitans :* Effects of sympathectomy and decerebration. *Zeitschr. f. Tierpsychol. 21,* 813-821.

Chernetski, K. 1965. Personal communication.

Chung, C., Witkop, C. and Henry, J. 1964. A genetic study of dental caries with special reference to PTC taste sensitivity. *Am. J. Human Genetics 16,* 231-245.

Clark, M. and Rand, M. 1964. A pharmacological effect of tobacco smoke. *Nature 201,* 507-508.

Corlis, R., Splaver, G., Wisecup, P. and Fischer, R. 1967. Myers-Briggs Type Personality Scales and their relation to taste acuity. *Nature 216,* 91-92.

Cox, G., Nathans, J. and Vonau, N. 1955. Subthreshold-to-taste thresholds of sodium, potassium, calcium and magnesium ions in water. *J. Applied Phys. 8,* 283-286.

Dawson, W., West, G. and Kalmus, H. 1967. Taste polymorphism to anetholtrithione and phenylthiocarbamate. *Ann. Hum. Genet. 30,* 273-276.

Duncan, C. 1963. Excitatory mechanisms in chemo- and mechanoreceptors. *J. Theoret. Biol. 5,* 114-126.

Dzendolet, E. 1969. Basis for taste quality in man, in : C. Pfaffmann, ed., Olfaction and Taste, Vol. 3, 420-427, Rockefeller Univ. Press, New York.

Eckert, L. 1969. Zur Wellung des Humanhaares. *J. Soc. Cosmetic Chemists 20,* 321-331.

Erickson, R., Doetsch, G. and Marshall, D. 1965. The gustatory neural response function. *J. Gen. Physiol. 49,* 247-263.

Erlenmeyer, H. 1932. As quoted in H. Friedman, Influence of isosteric replacements upon biological activity. Chemical-Biological Correlation, 296-306, 1951.

Eysenck, H. 1963. Experiments with Drugs, Pergamon/McMillan, London/New York.

Eysenck, H. 1965. Smoking, Health and Personality, Basic Books, New York.

Eysenck, H. and Eysenck, S. 1967. On the unitary nature of extraversion. *Acta Physiol. 26,* 282-290.

Eysenck, H. and Eysenck, S. 1970. Personality Structure and Measurement, Routledge and Kegan Paul.

Ferguson, J. 1939. The use of chemical potentials as indices of toxicity. *Proc. Roy. Soc. Biol. 27,* 387-404.

Filshie, B. and Rogers, G. 1961. The fine structure of α-keratin. *J. Mol. Biol. 3,* 784-786.

Fischer, R. 1953. The selectivity of Gram-stain for keratins. *Exper. 9,* 20-22.

Fischer, R. 1956. Affinity for wool as an indicator of neuropharmacological activity, 63-80, in: Grenell, R. and Mullins, L. eds., Molecular Structure and Functional Activity of Nerve Cells, Am. Inst. Biol. Sci.

Fischer, R. 1960. Selection on the molecular level : Biological role of drug-protein interaction, in : J. Segal, ed. Struktur und Biologische Funktion der Eiweisse, 165-174, 3rd Humboldt-Symposium über Grundfragen der Biologie Berlin, vom 21. September.

Fischer, R. 1966. Sex, lifespan and smoking. *Exper. 22,* 178-182.

Fischer, R. 1967 a. Genetics and gustatory chemoreception in man and other primates, in : M. Kare, ed., The Chemical Senses and Nutrition, 61-71.

Fischer, R. 1967 b. In a discussion following R. Henkin : Abnormalities of taste and olfaction in various disease states, in : M. Kare, ed., The Chemical Senses and Nutrition, 110-111.

Fischer, R. 1968. On creative, psychotic and ecstatic states, in : I. Jakab, ed., Psychiatry and Art, 117-146, Proc. 5th Int. Colloquium of the Soc. Psychopathol. Expression, Los Angeles, 1968, S. Karger, Basel and New York.

Fischer, R. 1969 a. Letter to the editor. *Psychophysiol. 5,* 591.

Fischer, R. 1969 b. Out on a phantom limb. Variations on the theme : Stability of body image and the golden section. *Persp. Biol. Med. 12,* 259-273.

Fischer, R. 1969 c. Letter. *Science 163,* 1144.

Fischer, R. 1970. Prediction and measurement of hallucinogenic drug-induced perceptual-behavioral change, in : W, Keup, ed., Proc. Ann. Meeting Eastern Psychiat. Assoc., New York City, Nov. 14-15, 1969, in press, New York.

Fischer, R. and Griffin, F. 1959. On factors involved in the mechanism of "taste-blindness" *Exper. 15,* 447-452.

Fischer, R. and Griffin, F. 1961 a. Biochemical-genetic factors of taste-polymorphism, Proc. 3rd World Congr. Psychiat., June 4-10, 1961, Montreal, Vol. 1, 542-547, Univ. of Toronto, 1. McGill U. Press.

Fischer, R. and Griffin, F. 1961 b. Taste-blindness and variations in taste threshold in relation to thyroid metabolism in normal and mentally ill populations. *J. Neuropsychiat. 3,* 98-104.

Fischer, R. and Griffin, F. 1962. Biochemical-genetic factors in health and mental retardation, Proc. 3rd World Congr. Psychiat. 1, 542-547.

Fischer, R. and Griffin, F. 1963. Quinine dimorphism : A cardinal determinant of taste sensitivity. *Nature 200,* 343-347.

Fischer, R. and Griffin, F. 1964 a. Pharmacogenetic aspects of gustation. *Arzneim.-Forsch. 14,* 673-686.

Fischer, R. and Griffin, F. 1964 b. Chemoreception and gustatory memory formation, Proc. 6th Int. Cong. Biochem. 8, 651.

Fischer, R. and Kaelbling, R. 1967. Increase in taste acuity with sympathetic stimulation: The relation of a just-noticeable taste difference to systemic psychotropic drug dose, in : J. Wortis, ed., Recent Advances in Biological Psychiatry 9, 183-195.

Fischer, R. and Larose, P. 1952 a. Contribution on the behavior and structure of the cytoplasmic membrane of bacteria. *Canad. J. Med. Sc. 30,* 86-105.

Fischer, R. and Larose, P. 1952 b. Mechanism of Gram-stain reversal. *J. Bact. 64,* 435-441.

Fischer, R. and Rockey, M. 1967. A steady-state concept of evolution, learning, perception, hallucination and dreaming. *Intl. J. Neurol. 6,* 182-201.

Fischer R. and Rockey, M. 1968. Psychophysics of excitation and tranquilization from a steady-state perspective. *Neurosci. Res. 1,* 263-314.

Fischer, R. and Seidenberg, S. 1951. Homologous mechanism of bactericidal action and Gram-staining. *Science 114,* 265-266.

Fischer, R. and Warshay, D. 1968. Psilocybin-induced autonomic, perceptual and behavioral change. *Pharmakopsychiatrie Neuro-Psychopharmakologie* (Thieme, Stuttgart) *1,* 291-302.

Fischer, R., Dunbar, H. and Sollberger, A. 1970. Are taste and drug sensitivity subject to periodic fluctuation? *Arzneim.-Forsch.,* in press.

Fischer, R., Griffin, F., England, S. and Garn, S. 1961a.Taste thresholds and food dislikes. *Nature 191,* 1328.

Fischer, R., Griffin, F., England, S. and Pasamanick, B. 1961 b. Biochemical-genetic factors of taste polymorphism and their relation to salivary thyroid metabolism in health and mental retardation. *Med. Exp. 4,* 356-366.

Fischer, R., Griffin, F. and Mead, E. 1962. Two characteristics of taste sensitivity. *Med. Exp. 6,* 177-182.

Fischer, R., Griffin, F. and Kaplan, A. 1963. Taste thresholds, cigarette smoking, and food dislikes. *Med. Exp. 9,* 151-168.

Fischer, R., Griffin, F. and Pasamanick, B. 1965 a. The perception of taste : Some psychophysiological, pathophysiological, pharmacological and clinical aspects, in : P. Hock and J. Zubin, eds., Psychopathology of Perception, 129-163, Grune and Stratton, New York, London.

Fischer, R., Griffin, F., Archer, R., Zinsmeister, S. and Jastram, P. 1965 b. Weber ratio in gustatory chemoreception; an indicator of systemic (drug) reactivity. *Nature 207,* 1049-1053.

Fischer, R., Griffin, F. and Rockey, M. 1966. Gustatory chemoreception in man : multidisciplinary aspects and perspectives. *Persp. Biol. Med. 9,* 549-577.

Fischer, R., Hill, R. and Warshay, D. 1969. Effects of the psychodysleptic drug psilocybin on visual perception : Changes in brightness preference. *Exper. 25,* 166-169.

Fischer, R., Hoelle, A. and Seidenberg, S. 1951. Voraussage der bakteriziden Wirkung von Substanzen durch Bestimmung ihrer Affinität zu Wolle.

Fischer, R., Kappeler, T., Wisecup, P. and Thatcher, K. 1970. Personality trait-dependent performance under psilocybin, Part 2, *Dis. Nerv. Sys. 31,* 181-192.

Fischer, R., Knopp, W. and Griffin, F. 1965. Taste sensitivity and the appearance of phenothiazine-tranquilizer induced extrapyramidal symptoms. *Drug Research 15,* 1379-1382.

Fischer, R., Marks, P., Hill, R. and Rockey, M. 1968. Personality structure as the main determinant of drug-induced (model) psychoses. *Nature 218,* 296-298.

Fischer, R., Ristine, L. and Wisecup, P. 1969. Increase in gustatory acuity and hyperarousal in schizophrenia. *Bio. Psychiat. 1,* 209-218.

Fischer, R., Thatcher, K., Kappeler, T. and Wisecup, P. 1969. Unity and covariance of perception and behavior. *Arzneim.-Forsch. 19,* 1941-1945.

Fisher, Sir Ronald. 1951. The Design of Experiments, p. 11, Hafner, New York.

Fox, A. 1932. The relation between chemical constitution and taste. *Proc. Nat. Acad. Sci. 18,* 115-120.

Freud, S. 1954. Das Unbehagen in der Kultur. *Gesammelte Schriften 12,* 27-114, Intern. Psychoanal., Wien.

Gale, A. and Coles, M. 1969. Brain waves and personality. *New Scientist 45,* 17-19.

Galton, F. 1968. As quoted in W. McCartney, Olfaction and Odours, p. 189, Springer-Verlag, Berlin and New York.

Gander, J., Griffin, F. and Fischer, R. 1964. A multiple-site chemoreceptor model. *Arch. Int. Pharmacodyn. 151,* 540-551.

Girgis, M. 1969. The amygdala and the sense of smell. *Acta Anat. 72,* 502-519.

Goedde, H. and Ohligmacher, H. 1965. Zur Problematik des Polymorphismus des Bitterschmeckens: Vergleichende Untersuchungen and Thioharnstoffderivaten und Anetholthrithion. *Humangenetik 1,* 423-436.

Goodman, L. and Gilman, A. 1941. The Pharmacological Basis of Therapeutics, MacMillan, New York.

Gordon, H. and Sperry, R. 1969. Lateralization of olfactory perception in the surgically separated hemispheres of man. *Neuropsychol. 7,* 111-120.

Graziadei, P. 1969. The ultrastructure of vertebrate taste buds, in : C. Pfaffmann, ed., Olfaction and Taste, Vol. 3, 315-330, Rockefeller Univ. Press, New York.

Griffin, F. 1966. On the interaction of chemical stimuli with taste receptors. Dissertation, The Ohio State University.

Guttman, R., Rosenzweig, K. and Guttman, L. 1965. The retest reliability of certain traits. *Acta Genet. 15,* 358-370.

Harris, H. 1955. An introduction to Human Biochemical Genetics, 69-75, Cambridge University Press, London.

Harris, H. and Kalmus, H. 1949. The measurement of taste sensitivity to phenylthiourea (PTC). *Ann. Eugen. Lond. 15,* 24-31.

Hartmann, G. 1939. Applications of individual taste difference towards phenylthiocarbamide in genetic investigations. *Ann. Eugen. 9,* 123-135.

Haukaas, H., Schor, R. and David, C. 1969. Statistical-mechanical studies of the a-β transformation in keratin. *Biophys. J. 9,* 1252-1255.

Hecht, K., Choinowski, S., Solle, M. and Treptow, K. 1961. Die Bedeutung der individuellen Erregbarkeit des ZNS für den Effekt zentralwirkender Pharmaka. *Acta Physiol. 20,* 119-134.

Hecht, K., Krause, H., Misgeld, G. and Treptow, K. 1960. Zur Differenzierung von Typen der höheren Nerventätigkeit bei Ratten. *Acta Biol. Med. Germ. 4,* 254-268.

Hecht, K. and Treptow, K. 1961. Zur Frage des "Typus" der höheren Nerventätigkeit bei Albinoratten. *Acta Physiol. 20,* 103-117.

Hess, E. 1965. Attitude and pupil size. *Sci. Am. 212,* 46-54.

Hess, E. and Polt, J. 1966. Changes in pupil size as a measure of taste difference. *Percept. and Motor Skills 23,* 451-455.

Hijmans, A. 1962. The fringe of pharmacology : Phytopharmacology. *Acta Physiol. Pharmacol. Neerlandica 11,* 98-103.

Horne, R. 1970. Water : Nomenclature. *Science 168,* 151.

Jellinek, G. 1964. Modern methods of sensory analysis. *J. Nutr. Diet. 1,* 219-260.

Jörgensen, G. 1969. Schmecken und Riechen, in : P. Becker, ed. Humangenetik, p. 108, Georg Thieme Verlag, Stuttgart.

Joyce, C., Pann, L. and Varonos, D. 1968. Taste sensitivity may be used to predict pharmacological effects. *Life Sciences 7,* 533-537.

Kalmus, H. and Hubbard, S. 1960. The Chemical Senses in Health and Disease, Charles Thomas, Springfield, Illinois.

Kaplan, A. and Fischer, R. 1965. Taste sensitivity for bitterness : Some biological and clinical application, in : J. Wortis, ed., Proc. 19th Ann. Meeting Soc. Biol. Psychiat. 183-196.

Kaplan, A., Fischer, R., Karras, A., Griffin, F., Powell, W., Marsters, R. and Glanville, E. 1967. Taste thresholds in twins and siblings. *Acta Genet. Med. Gemellol. 16,* 229-243.

Kaplan, A., Glanville, E. and Fischer, R. 1964. Taste thresholds for bitterness and cigarette smoking. *Nature 202,* 1366.

Kaplan, A., Glanville, E. and Fischer, R. 1965. Cumulative effect of age and smoking on taste sensitivity in males and females. *J. Gerontol. 20,* 335-337.

Kare, M. 1969. Digestive functions of taste stimuli, in. C. Pfaffmann, ed., Olfaction and Taste, Vol. 3, 586-592, The Rockefeller University Press, New York.

Kay, C. 1969. Smoking habits of oral contraceptive users. *Lancet, ii,* 1228-1229.

Kimura, K. 1961, Factors affecting the response of taste receptors of the rat. *Kumamoto Med. J. 14,* 95-99.

Kimura, M., Waki, I. and Ikeda, H. 1968. Potentiation of the crude drug "moschus" for catecholamines. *Yakugaku Zasshi 88,* 130-134.

Klein, T. and DeFries, J. 1970. Similar polymorphism of taste sensitivity to PTC in mice and men. *Nature 225,* 555-557.

Krut, L., Perrin, M. and Bronte-Stewart, B. 1961. Taste perception in smokers and non-smokers. *Brit. Med. J. 1,* 384-387.

Kubota, T. and Kubo, I. 1969. Bitterness and chemical structure. *Nature 223,* 97-99.

Langecker, H. 1969. Geschlechtdifferenzen von Enzymaktivitäten in verschiedenen Organen. *Arzneim.-Forsch. 19,* 1769-1776.

Larose, P. and Fischer, R. 1953. Reversal of Gram-staining behavior. *Science 117*, 449.

Läuger, P., Martin, H. and Müller, P. 1944. Ueber Konstitution und toxische Wirkung von natür-
lichen und neuen synthetischen insektentötenden Stoffen. *Helv. Chim. Acta 27*, 892-
928.

Lemberger, F. 1908. Psychophysische Untersuchungen über den Geschmack von Zucker und
Saccharin. *Arch. Ges. Physiol. 123*, 293-311.

Lotan, N., Yaron, A. and Berger, A. 1966. The stabilization of the *a*-helix in aqueous solution
by hydrophobic side-chain interaction. *Biopolymers 4*, 365-368.

Lucretius, T. 1954. As quoted by R. Hainer, A. Emslie and A. Jacobson : An information theory of olfaction, in : R. Miner, ed., "Basic Odor Research Correlation", *Ann. N.Y. Acad.
Sci. 58*, 158-174.

McGlothlin, W. and Cohen, S. 1965. The use of hallucinogenic drugs among college students. *Amer.
J. Psychiat. 122*, 572-574.

MacKinnon, D. 1966. The nature and nurture of creative talent, in : B. Semeonoff, ed., Personality Assessment, Penguin Books, Baltimore.

MacKinnon, D. 1967. Personal communication.

MacLean, P. 1959. The limbic system with respect to two basic life principles, in : M. Brazier, ed.,
The Central Nervous System and Behavior, 31-118, Josiah Macy Jr., Foundation, New
York.

MacLean, P. 1962. New findings relevant to the evolution of psychosexual functions of the brain.
J. Nerv. Ment. Dis. 135, 289-301.

Mazur, R., Schlatter, J. and Goldkamp, A. 1969. Structure-taste relationships of some dipeptides.
J. Am. Chem. Soc. 91, 2684-2691.

Merton, B. 1958. Taste sensitivity to PTC in 60 Norwegian families with 176 children. *Acta Genet.
8*, 114-128.

Meyer, K., Mark, H. and Astbury, W. 1961. See p. 180 in : E. Mercer, Keratin and Keratinization,
Pergamon, New York, London.

Michael, R. and Keverne, E. 1970. Primate sex pheromones of vaginal origin. *Nature 225*, 84-85.

Mitoma, C. 1970. As quoted in G. Cosmides' report of a Meeting of the N.I.H. Pharmacology-To-
xicology Branch. *Science 168*, 1013.

Morrison, R. and Ludvigson, H. 1970. Discrimination by rats of nonspecific odors of reward and
nonreward. *Science 167*, 904-905.

Mozell, M., Smith, B.P., Smith, P.E., Sullivan, R.L. Jr., and Swender, P. 1969. Nasal chemoreception in flavor identification. *Arch. Otolaryng. 90*, 367-373.

Murphy, M. and Schneider, G. 1970. Olfactor bulb removal eliminates mating behavior in the male
golden hamster. *Science 167*, 302-303.

Nauta, W. 1960. Anatomical relationships between the amygdaloid complex, the dorso-medial tha-
lamic nucleus and the orbito-frontal cortex in monkey (abstract), *Anat. Rec. 136*, 251.

Oakley, B. 1967. Altered taste responses from cross-regenerated taste nerves in the rat, in : T.
Hayashi, ed., Olfaction and Taste, Vol. 2, 535-547, Pergamon, New York/London.

Oakley, B. 1969. Taste preference following cross-innervation of rat fungiform taste buds. *Physiol. and Behav. 4*, 929-933.

Parry, D. 1969. An alternative to the coiled-coil for α-fibrous proteins. *J. Theoret. Biol. 24*, 73-84.

Parry, D. 1970. A proposed conformation for α-fibrous proteins. *J. Theor. Biol. 26*, 429-435.

Pauling, L. and Corey, R. 1951. The pleated sheet, a new layer configuration of polypeptide chains. *Proc. Natl. Acad. Sci. 37*, 251-256.

Pauling, L., Corey, R. and Branson, H. 1951. The structure of proteins : Two hydrogen-bonded helical configurations of the polypeptide chain. *Proc. Natl. Acad. Sci. 37*, 205-211.

Perutz, M., Bolton, W., Diamond, R., Muirhead, H. and Watson, H. 1964. Structure of hemoglobin. *Nature 203*, 687-690.

Pfaffmann, C. 1962. On the code for gustatory sensory quality, in : R. Gerard and J. Duyff, eds., Information Processing in the Nervous System, Vol. 3, 267-273, Excerpta Medica Found., New York/London.

Pfaffmann, C. 1969. Summary of taste roundtable, in : C. Pfaffmann, ed., Olfaction and Taste, Vol. 3, 527-532, Rockefeller Univ. Press, New York.

Reis, D., Carmichael, M. and MacLean, P. Cerebral representation of genital function. IV : Frontotemporal region (unpublished).

Robbins, N. 1967. "Electric taste" after section of the chorda tympani. *Nature 214*, 1113-1114.

Rubin, T., Griffin, F. and Fischer, R. 1962. A physico-chemical treatment of taste thresholds. *Nature 195*, 362-364.

Sato, M. 1963. The effect of temperature change on the response of taste receptors, in : Y. Zotterman, ed., Olfaction and Taste, Vol. 1. 151-164, Pergamon, New York.

Schönpflug, W. 1969. Magnitude, judgment and arousal, as quoted in G. Benedetti, Das Unbewusste in neuropsychologischer Sicht. *Der Nervenarzt 40*, 149-155.

Schutz, H. and Pilgrim, F. 1957. Differential sensitivity in gustation. *J. Exp. Psychol. 54*, 41-48.

Shallenberger, R. and Acree, T. 1967. Molecular theory of sweet taste. *Nature 216*, 480-482.

v. Skramlik, E. 1963. The fundamental substrates of taste, in : Y. Zotterman, ed., Olfaction and Taste, Vol. 1, 125-132, Pergamon, New York/London.

Skude, G. 1963. Some factors influencing taste perception for phenylthiourea (PTC). *Hereditas 50*, 203-210.

Smith, W., Powell, E. and Ross, S. 1955. Manifest anxiety and food aversions. *J. Abnormal Soc. Psychol. 50*, 101-104.

Speakman, J. 1936. Cross-linkage formation in keratin. *Nature 138*, 327.

Stephan, H. : 1966. Grössenänderungen im olfaktorischen und limbischen System während der phylogenetischen Entwicklung der Primaten, in : R. Hassler and H. Stephan, eds., Evolution of the Forebrain, 377-386, Georg Thieme Verlag, Stuttgart.

Stigter, D. 1964. Ionic conduction and liquid flow in wet wool. *J. Colloid Sci. 19*, 252-267.

Stone, H. and Pangborn, R. 1968. Intercorrelation of the senses, in : Basic Principles of Sensory Evaluation, Special Technical Publication No. 433, Am. Soc. Testing and Materials.

Straus, E.W. 1965. Born to see, bound to behold, *Tijdschrift Voor Filosofie 27e*, 659-688, quoting Ovidius Publius Naso : *Metamorphoses 1*, 84-86.

Tarriére, C. and Hartemann, F. 1964. Investigation into the effects of tobacco smoke on a visual vigilance task. Ergonomics, Proc. 2nd I.E.A. Congress, Dortmund, 525-530.

Teranishi, R. 1970. Private communication with G. Ohloff.

Thatcher, K., Kappeler, T., Wisecup, P. and Fischer, R. 1970. Personality trait-dependent performance under psilocybin. *Dis. Nerv. Sys. 31*, 181-192.

Thatcher, K., Wiederholt, W. and Fischer, R. To be published.

Thomas, C. and Cohen, B. 1960. Comparison of smokers and nonsmokers. *Bull. Johns Hopkins Hosp. 106*, 205-214.

Thompson, R., Mayers, K. and Patterson, C. 1970. Number-coding in the association cortex of the cat. *Science 168*, 271-273.

Trotter, W. 1962. The relative toxicity of antithyroid drugs. *J. New Drugs 2*, 333-343.

Vinnikov, J. 1965. Principles of structural, chemical and functional organization of sensory receptors. *Cold Spring Harbor Symposia on Quant. Biol. 30*, 293-299.

Wolfram, L. 1965. Reactivity of disulphide bonds in strained keratin. *Nature 206*, 304-305.

Wolfgram, F. and Rose, A. 1962. The amino acid composition of central and peripheral nerve neurokeratin. *J. Neurochemistry 9*, 623-627.

Wright, R. 1966 a. Why is an odor? *Nature 209*, 551-554.

Wright, R. 1966 b. Odor and molecular vibration. *Nature 209*, 571-573.

Yokotsuka, T., Saito, N., Okuhara, A. and Tanaka, T. 1969. Studies on the taste of a-amino acids : Part 1, Ternary synergism of palatable taste of glycine. *NOKA (Agric. Chem.) 43*, 165-170.

Zalewski, A. 1970. Regeneration of taste buds in the lingual epithelium after excision of the vallate papilla. *Exp. Neurol. 26*, 621-629.

DISCUSSION

Jones :   Does taste threshold predict a level of autonomic activity, like the Funkenstein test does?

Fischer :   Since taste thresholds reflect general, organismic sensitivity they are rather stable, in contrast to autonomic activity, which fluctuates. The high cortical arousal of acute schizophrenics represents a higher baseline of autonomic activity which further reduces the variability of such persons. No wonder that acute schizophrenics display more stable taste thresholds than normophrenics.

Hughes :   1 : Does the higher mean energy level among the sensitives refer to alpha activity of relatively high frequency or high amplitude? I would suppose it to be the former. One might also have a reference electrode intruded into the central vertex to check eye movements, since the corneal retinal potential gets over 200 mV. If the patients were allowed to blink as they wished, you might find the taste-sensitive subjects are blinkers and the others non-blinkers.

2. Is this evidence about dreaming based solely upon the answer to a questionnaire? The best way to do this would be to carry out full night sleep recordings and find who had a high percentage of rapid eye movement stage of sleep.

Fischer :

In order to avoid the difficulties to which you refer, we used a Drohocki integrator and thus measured only amplitudes. With increasing levels of arousal the variability of the EEG decreases. To answer your second question, my comment that taste sensitives dream more frequently and in color was based only on questionnaires, but we are beginning to study full-night dream records to determine whether the frequency of the REM state is greater in taste sensitives.

Beets :

We have carried out a panel test on specific anosmia effects with substances (e.g. steroid ketones) having an intense urine odor for most people. We found that about 1/3 of the population cannot smell this typical note. Among these people, there are trained perfumers, who are very sensitive to all types of odor. How does this agree with your generalization of sensitivity?

Fischer :

I do not know.

Kalmus :

If Dr. Beets is dealing with this specific anosmia, you would not expect these general conditions which Prof. Fischer mentioned to prevail.

Schudel :

Some perfumers have smoking habits that do not prevent them from being good perfumers. Although decreasing taste sensitivity, can smoking also increase the odor sensitivity?

Fischer :

Smoking, i.e., the pharmacological action of nicotine, is first exciting and then depressing; furthermore, some people smoke for the stimulating effect (they are the extrovert, insensitive tasters with low cortical arousal), but some smoke for other reasons. Remember that smell may act as a cue for taste. I do not know what kind of a personality type perfumers are, but I can only compare them to our dieticians, who are mostly "non-tasters". They really should not smoke, but if they smoke, they surely need it. Most likely some people smoke for the excitatory and others for the tranquillizing effect of smoking.

Kovats :

Is it true that a good perfumer is not primarily a sensitive person but someone who remembers odors well?

Fischer :

I don't know but if it is true then perfumers must be 'stable' people, because good memory is another aspect of stability. 'Variable' persons with large standard deviations on perceptual and behavioral tasks **re-experience** rather than remember. Maybe you do not need this as a perfumer, because it would interfere with your judgment. We are dealing here with both the sensitivity and the stability of performance, and with the pharmacological response to nicotine, as well.

Eykelboom :

How many times have you determined the stability of the threshold value and on how many people have you done it? Sometimes we find, looking at a small number of repetitions, some people tend to be rather stable, but if you look at a group of people, you find that nearly everybody is unstable.

Fischer : We use the handwriting area as a convenient test for stability in computing the standard deviations from four times copying a 28-word text, and we have done this with about 60 or 70 people up to 8-10 times. We test a subject three times in three weeks. If he shows a stable small standard deviation each time, then we are sure that he will always be stable on nearly any performance. If, however, a subject first shows a small standard deviation, and then a larger one a week later, and a medium-size one two weeks later, we know that he is a variable subject. The size of a subject's standard deviation predicts the degree of stability he will display on a perceptual-behavioral performance on that day.

Harper : Do the observations about quinine apply also to the responses to sucrose and to sodium chloride?

Fischer : Yes, they do. I used in the presentation only quinine and PROP because they are the archetypal representatives of Gaussian and bimodal compounds, respectively, and moreover, because quinine has a convenient range of solubility in water.

A ROUND TABLE DISCUSSION ON THE

CELLULAR-MOLECULAR BASIS OF CHEMICAL PERCEPTION

Chairman : M.R. Kare

Participants : Mrs. L.M. Riddiford, G. Kasang, G.J. Henning, R.H. Cagan

Kare :   Taste and smell are specialized receptors for communication with the chemicals in our environment. A living organism communicates with the chemistry of its environment in the essential activities of respiration, ingestion of food and fluid and, in many species, reproduction. Until now, most of the discussion at this symposium has isolated taste and smell from the rest of the body, as though they were distinct from and uninfluenced by body functions. We know that this is not the case. Environment, age, disease and nutritional state are some factors that modify olfactory and gustatory perceptions.

Our round table will discuss the cellular-molecular basis of perception. Rather than considering it in terms of the anatomy of the cells, we are going to examine these senses as function of a complex organism – approaching the subject from the point of view of the function of these senses in the body.

Our first two speakers (Henning and Cagan) will provide a general approach to the subject; the last two speakers (Kasang and Riddiford) are going to talk about specific organisms and specific receptors. We have agreed to consider the subject from the standpoint of ideas rather than data. Afterwards, we will discuss a few questions among ourselves, so that you can see the same question from different perspectives. Finally we will give you an opportunity to cross-examine us about some of these new ideas.

SOME GENERAL CRITERIA FOR THE RECEPTOR PROCESS IN

GUSTATION AND OLFACTION

G.J. Henning

Unilever Research, Vlaardingen,
The Netherlands.

Our knowledge about the nature of the receptor process in taste and olfaction derives mainly from psychophysical and electrophysiological experiments. Sensory cells generally transform the information presented to them into relatively slow changes in voltage across the receptor cell membrane, the so called receptor potentials. These potentials are graded according to the stimulus strength and in olfactory and gustatory cells are known to appear within a few 100 ms after stimulation. Furthermore, from calculations on the minimal number of molecules required to elicit an odor sensation (Stuiver, 1958) or a behavioral response in insects (Boeckh et al., 1965), it follows that olfactory cells can function as molecule counters, able to signal the number of molecular impacts per unit of time. Since the energy to be gained simply by the adsorption of a few molecules is much less than that required to generate the receptor potential, the latter obviously constitutes an amplified signal. The same holds in vision (Hagins, 1965; Cone, 1965), and probably in taste. Finally, regarding the discriminatory capacity of the receptor cells, electrophysiology has shown that for both taste and odor one receptor cell responds to many different stimuli.

From the above we can derive the following general criteria for the receptor process in olfaction and gustation :

– **the graded transduction of a chemical into an electrical signal;**

-- **some amplification should be allowed for;**

– **the process should be fast enough to generate a potential within a few 100 ms;**

– **the mechanism proposed must meet modest specificity requirements.**

Most of the existing theories on odor and taste satisfy only part of these criteria. More complete theories have been advanced by Duncan (1964) for taste and by Davies (1965) for olfaction.

Across the membrane of sensory cells, made up mainly of proteins, lipids, water and inorganic ions, a potential difference exists, which originates from an uneven distribution of diffusable ions over the fluids inside and outside the cell. This, called the **"resting potential"** (10-100 mV, negative inside) is maintained by metabolic processes. The receptor potential is usually considered as a transient change in the resting potential, resulting from a stimulus-induced change in permeability of the membrane to certain ions (K, Na, Cl). One other possibility would be that the receptor potential is caused by a change in activity of an ion-transporting enzyme system present in the membrane (e.g. a Na/K pump). If such an ion pump contributes directly to the membrane potential, by transporting unequal amounts of ions into and out of the cell, any change in pumping activity will change the transmembrane potential and generate a receptor potential (Fig. 1).

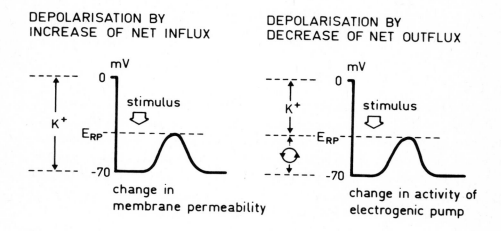

DEPOLARISATION BY
INCREASE OF NET INFLUX

DEPOLARISATION BY
DECREASE OF NET OUTFLUX

change in
membrane permeability

change in activity of
electrogenic pump

Fig. 1. Two possible mechanisms for the generation of a receptor potential. E_{RP}= equilibrium point of the receptor potential.

Such a process participates in the generation of various potentials, including the receptor potential in the eye of Limulus (Smith et al., 1968), and in the olfactory epithelium of the frog (Henning and Sierevogel-de Maar). Both generator processes are probably fast enough to comply with the short reaction times required, but it is conceivable that they both need additional amplification. Furthermore, the mere generation of a potential will not allow for much differentiation between stimuli, and more specific receptive structures will be needed.

In analogy with other chemoreceptive systems we may assume that the sites recognizing stimulus are on the outside of the membrane of the taste and olfactory cells. Which of the constituents of the cell membrane is the most likely candidate to function as the receptor? In nearly all cases where a living system responds to a chemical signal, the molecule recognizing the signal is a protein. Indeed, proteins seem to possess the necessary structural features to perform such a function, since by appropriate folding of their polypeptide chains, sufficiently rigid receptor sites with various shapes and having stereospecific recognition capacities can be formed. But, how could the process of stimulus recognition be coupled with the generation of a potential within a reasonable time? Perhaps the **"regulatory"** enzymes can serve as a model here, as has been suggested by Changeux et al. (1967).

It then has to be assumed that excitable membranes contain aggregates of lipoprotein units to which at least two conformational states are reversibly accessible (Changeux et al. 1968), and that the change in conformation is a cooperative affair, i.e. all lipoprotein units in an aggregate

REGULATORY ENZYME

RECEPTOR CELL MEMBRANE

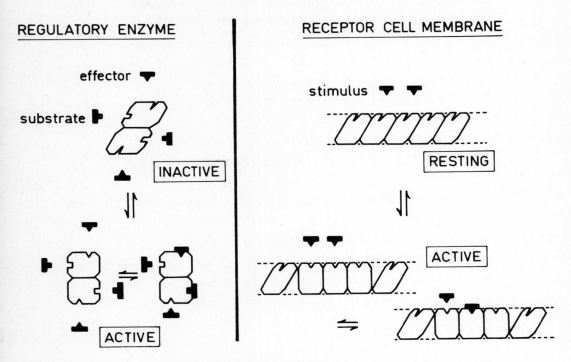

Fig. 2. Allosteric interactions in a regulatory enzyme and an excitable membrane.

change their conformation simultaneously (Fig. 2). When the two conformational states correspond with different permeabilities of the membrane to ions, or with a different activity of an ion pumping system in the membrane, the change from one conformation to the other will result in the generation of a potential.

The receptor process in taste and olfaction thus could be envisaged as the binding of the stimulus to regulatory sites on such lipoprotein assemblies, tipping the balance between the two conformational states by allosteric interaction and generating a potential. As in this case the stimulus binds to a protein, the geometry of the stimulus molecule can be expected to play a determinant role, thus accounting for the desired specificity of the cellular response. Furthermore, conformational changes do not involve the making or breaking of covalent bonds, so that the process will be fast enough to provide the speed of response required. Finally, depending on the number of lipoprotein aggregates "**coupled**", the necessary amplification can be achieved.

REFERENCES

Boeckh, J., Kaissling, K.E. and Schneider, D. 1965. Insect olfactory receptors. *Cold Spring Harbor Symp. Quant. Biol. 30,* 263-280.

Changeux, J.P., Thiéry, J., Tung, Y. and Kittel, C. 1967. On the cooperativity of biological membranes. *Proc. Nat. Acad. Sci. 57,* 335-341.

Changeux, J.P., and Thiéry, J. 1968. On the excitability and cooperativity of biological membranes, in : Regulatory Functions of Biological Membranes, (J. Järnefelt, Ed.) BBA Library Vol. 11, Elsevier Publ. Cy., Amsterdam, p. 116-138.

Cone, R.A. 1965. The early receptor potential of the vertebrate eye. *Cold Spring Harbor Symp. Quant. Biol. 30,* 483-491.

Davies, J.T. 1965. A theory of the quality of odors. *J. Theoret. Biol. 8,* 1-7.
        1969. The "penetration and puncturing" theory of odor : types and intensities of odors. *J. Colloid Interface Sci. 29,* 296-304.

Duncan, C.J. 1964. The transducer mechanism of sense organs. *Naturwiss. 51,* 172-173.
        1965. Cation-permeability control and depolarization in excitable cells. *J. Theoret. Biol. 8,* 403-418.

Hagins, W.A. 1965. Electrical signs of information flow in photoreceptors. *Cold Spring Harbor Symp. Quant. Biol. 30,* 403-418.

Henning, G.J., Sierevogel-de Maar, H.J.J.A. To be published.

Smith, T.G., Stell, W.K., Brown, J.E., Freeman, J.A. and Murray, G.C. 1968. A role for the sodium pump in photoreception in Limulus. *Science 162,* 456-458.

Stuiver, M. 1958. Biophysics of the sense of smell. Doctoral Thesis State University Groningen, The Netherlands.

COUNTERPARTS OF GUSTATORY RECEPTORS

Robert H. Cagan

Veterans Administration Hospital and Monell Chemical Senses Center,
University of Pennsylvania, Philadelphia, Pennsylvania (U.S.A.).

Possible counterparts of gustatory receptors present themselves in enzymes and in membrane transport proteins. Many features of these molecules provide a useful conceptual framework for approaching the problem of taste receptor molecules. In this context, enzymes should be considered not as catalysts, but as "**recognition sites**" for particular molecules. Similarly, the recently-discovered membrane transport proteins (New York Heart Association, 1969) interact very specifically with particular substrates, and are also therefore recognition sites.

Several analogies may be drawn among the three types of "**recognizer molecules**" – on the one hand enzymes and membrane transport proteins and on the other, taste receptor molecules. Enzymes are proteins that cover a wide range of molecular weight, from about 10,000 up to very complex enzyme systems of several million. Commonly the larger ones are composed of several subunits. Membrane transport proteins are of molecular weight around 30,000 (Pardee, 1968), and appear to be intimately associated with the plasma membrane. The catalytic function of enzymes is so well-documented that it is a truism of modern biology. The mechanism of action of transport proteins, however, is not known. Although not a membrane transport protein, the presumed acetylcholine-receptor protein can be considered in the same context.

Much is now known of the active site of many enzymes. Only indirect evidence is available on the chemistry of the binding site of membrane transport proteins, but the experimental evidence also points to highly specific binding sites. Taste receptor molecules are the ones about which least is known. Any attempts at isolation should be cognizant of the physiological taste data on specificity in the same species, such as relative "**sweetness**" of carbohydrates or relative "**bitterness**" (Dastoli and Price, 1966), but compare (Dastoli et al. 1968; Price, 1969). The existence of receptor molecules with the capacity to recognize taste stimuli remains a working hypothesis only. The analogy between the specificity of a taste receptor recognition site and an active site of an enzyme should not be drawn too closely. Relative specificity varies considerably among different enzymes. The question of specificity in a general sensor such as a taste receptor is and should remain open until experimental evidence is presented.

The role of the structure of the stimulus has received considerable attention (Shallenberger, 1970). In the field of enzymology, study of the substrate itself, while necessary, has never proven to be sufficient for defining the nature of the active site. It may be expected that the same will hold true for chemoreceptors.

The binding constant of a particular type of site for its substrate should be applicable to taste receptor molecules, as it has been for enzymes and membrane transport proteins. The capacity of the system, i.e., the number of recognition sites in the taste receptor field, should be considered at the same time. It is also possible that differences in specificity lie in differing efficiencies of receptor molecules in energy transduction, although such a function of the receptor molecule itself can only be speculated upon at this point. Phenomena that have been well-demonstrated in enzymes such as conformational changes, cooperativity, and allostery have not been looked for in detail yet in most membrane transport proteins. In one recent case that has been studied, the evidence showed that although a leucine-binding protein is able to undergo reversible conformational changes, the substrate leucine does not cause such changes (Penrose et al., 1970). In the case of taste receptor molecules current knowledge is even more primitive.

REFERENCES

Dastoli, F.R. and Price, S. 1966. Sweet-sensitive Protein from Bovine Taste Buds. Isolation and Assay. *Science 154,* 905.

Dastoli, F.R., Lopiekes, D.V. and Doig, A.R. 1968. Bitter-sensitive Protein from Porcine Taste Buds. *Nature 218,* 884.

New York Heart Association 1969. Proceedings of Symposium on "Membrane Proteins", Little, Brown and Co., Boston.

Pardee, A.B. 1968. Membrane Transport Proteins. *Science 162,* 632.

Penrose, W.R., Zand, R. and Oxender, D.L. 1970. Reversible Conformational in a Lenzine-Binding Protein from Escherichia Coli. *J. Biol. Chem. 245,* 1432.

Price, S. 1969. Putative Bitter-taste Receptor from Porcine Tongues. *Nature 221,* 779.

Shallenberger, R.S. 1970. Internat. Symposium on Gustation and Olfaction, Geneva. This volume.

BOMBYKOL RECEPTION AND METABOLISM ON THE ANTENNAE

OF THE SILKMOTH BOMBYX MORI

Gerhard Kasang

Max-Planck-Institut für Verhaltensphysiologie,
Seewiesen, und
Max-Planck-Institut für Biochemie, München.

Specific interaction between a chemoreceptor cell and the stimulating molecule requires that either one or both of the interaction partners have specific physico-chemical properties. What is the mechanism of the odor molecule – "**acceptor molecule**" – interaction, and how is the cell eventually switched into the excited state as seen from the bioelectric response? Are the odor molecules during or after the interaction with the acceptor chemically transformed?

In order to study such interactions between pheromones and insect olfactory organs, we have synthetized bombykol in a highly tritiated form (fig. 1) (Kasang, 1968). Naturally occurring bombykol is a long-chain unsaturated alcohol. It was isolated and analyzed by Butenandt et al. (1961) and used by our group for electro-physiological investigations of the receptors on the male moth's antenna (Schneider, 1970). Bombykol receptors have been carefully examined electron microscopically (Steinbrecht, 1970).

specific radioactivity	formula	rel. biol. activity (EAG)
125.6 mCi/mg =30.2 Ci/mmol	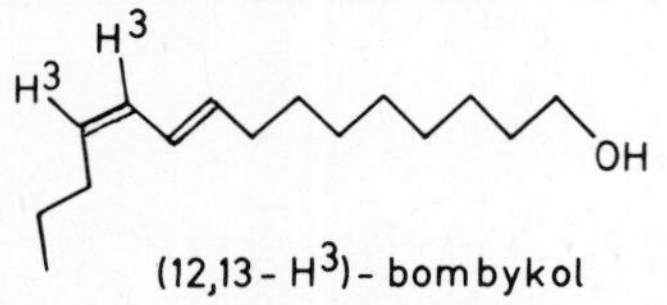 H^3 H^3 ⌇⌇⌇OH (12,13-H^3)- bombykol	1
250.1 mCi/mg =60.4 Ci/mmol	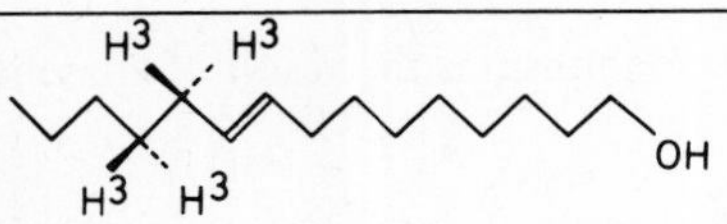 H^3 H^3 H^3 H^3 ⌇⌇⌇OH (12,13-H^3)- dihydrobombykol	10^{-3}

detection limit with liquid scintillation counter: 10^8 molecules

Fig. 1.

Not only the normal, but also the tritiated bombykol with one tritium atom per molecule, shows high biological activity in behavior and electro-physiological tests (Kaissling and Priesner). Using liquid scintillation counters, as few as 10^8 labeled molecules can be detected. The high tritium titer of our bombykol permits us to study the bombykol reception under physiological conditions. Tritiated bombykol undergoes metabolic transformation on the antenna. The amount of material extracted from the antenna – at different time intervals after bombykol application – was too small for direct chemical analysis. Nevertheless, chromatographic comparison of the eluted material with 100 substances of the lipid-class permits us to say that the bombykol metabolites are fatty acids, fatty acid-esters, and fatty alcohols (fig. 2).

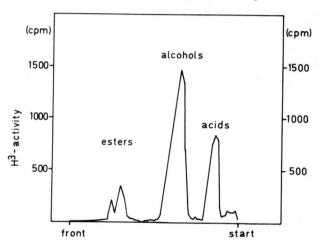

Fig. 2.

Apparently, the bombykol metabolism is the result of enzymatic activity, for the following reasons :

1. The metabolites are not the result of autoxidation, polymerization, autoradiolysis, or reaction with antennal lipids.

2. The metabolism depends upon the pH and the temperature.

3. The transformation of bombykol into the metabolites is completely inhibited by 2,4-dichlorophenol and other denaturating substances (fig. 3).

Inhibition of H^3 - Bombykol Metabolism

metabolism	complete inhibition	partial inhibition	activation
acids	2,4 - dichloro - phenol 10^{-2} molar	a) iodoso-benzoat and iodacetate 10^{-2} molar b) p-chloro-mercuri-benzoat 10^{-3} molar c) dodecylsulfate 10^{-2} molar	a) potassium cyanide 10^{-2} molar b) 1,10-phenan-throline 10^{-2} molar c) formaldehyde 10^{-2} molar
esters	2,4 - dichloro - phenol 10^{-2} molar		

buffer: 10^{-1} molar tris or phosphate

Fig. 3.

The transformation of bombykol into fatty esters and fatty acids seems to be the result of the activity of two different enzyme systems, because of the following facts.

1. After hydrolysis of the ester, radioactivity is only found in the alcoholic, but not in the fatty acid fraction.
2. Only the transformation of bombykol into the acid is blocked by sulfhydryl-inhibitors and by detergents, but activated by chelating agents.
3. Neither sulfhydryl-inhibitors, nor the detergents block the formation of the ester.

In order to answer the question of whether it is possible to localize the two metabolic processes and the receptor-bombykol interaction on the antenna, we used fractionated elution of the antennae with n-pentane and a mixture of chloroform and methanol after the application of bombykol (fig. 4).

It appeared that the amount of metabolites which could be eluted depends upon the incubation time and the solvent used. After an incubation of up to 2 minutes, the bulk of radioactivity is eluted by n-pentane and largely present in the fatty alcohols. With extended incubation times and elution by chloroform-methanol, the radioactivity is mainly found in the ester and acid fractions. So far, we are not able to say whether the eluted alcohols are unchanged or isomerized bombykol. A small amount of radioactive "**residual**" activity can only be recovered after an extensive elution of the antennae with polar solvents. This residual activity is composed of strongly polar, acidic compounds which seem to have undergone relatively strong binding to the antennal tissue. The speed of the metabolic bombykol transformation is high soon after the application, decreases rapidly, and reaches zero after about 30 minutes. By this time, approximately 80 per cent of the bombykol is transformed.

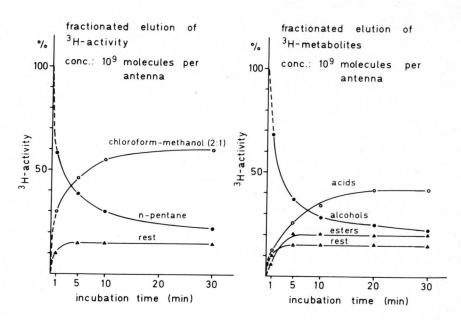

Fig. 4.

How can we interpret these observations?

1. It is reasonable to assume that the bombykol is first adsorbed on the antenna, then diffused from an outer lipid phase into an inner aqueous phase, namely the sensillum liquor and the cytoplasm of the receptor dentrite. In the tissue elements, that are not chemosensory, the analogous sequence would be : cuticle adsorption and cuticle penetration, diffusion into the epidermal cells, and possibly also into the hemolymph (fig. 5.)

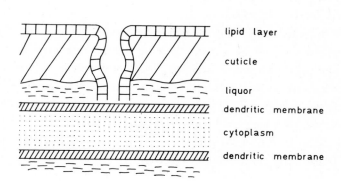

Fig. 5.

2. Since the bombykol-transforming enzymes are competing with one another, one may assume that they occur in the same locality, for instance in one and the same sensory hair.

3. The fact that after 30 minutes of incubation only 80 per cent of the alcohol has been transformed could be the result of an equilibrium process, a competitive inhibition by reaction-products.

4. Bombykol transformation into the metabolites and bombykol transfer (migration) from a lipid phase into an aqueous one are independent processes.

Apparently, the metabolization of bombykol into fatty acids and esters is not related to the excitation of the bombykol receptor cell, for the following reasons :

1. Bombykol is also transformed on female antennae which do not respond to bombykol.

2. Other *Bombyx* tissues are also capable of transforming bombykol into similar acid metabolites.

3. Labeled dihydro-bombykol shows a similar metabolism, but has an 10^3 times reduced olfactory effect.

4. The olfactory cells have a reaction time of milliseconds, but the metabolites can only be detected some seconds after the bombykol application.

It is now necessary to consider the possible biological meaning of the observed bombykol metabolization. There are two functions one can think of. Both are related to the only known role bombykol plays in the life of this species. This is the stimulation of the male's bombykol receptors during the time when the female is expanding her lure glands and ready for copulation.

1. Any bombykol molecule which adsorbs during the luring phase somewhere on the female's or male's body, and not the male's bombykol receptor hairs, is misplaced and may later desorb when the female is no longer receptive. These desorbed molecules could now stimulate the male at the wrong moment and therefore convey information that is now unwanted and false. The enzymatic bombykol transformation found in all the *Bombyx* tissues which were studied so far, takes care of this problem.

2. The excitation of the bombykol receptor cell is the result of its interaction with this pheromone. Presumably this interaction leads to a conformational change of the hypothetical bombykol acceptors in the dendritic membrane. This process may not involve a change of the hydroxyl group of the bombykol. It would now be the task of the metabolizing enzymes to break down the bombykol, again in order to avoid a re-stimulation of the cell and an unwanted information transfer.

Finally, the experiments leave the possibility open that bombykol is isomerized during or following its interaction with the receptor membrane, Isomerization would require a specific enzyme and would,of course, be a chemical process.

REFERENCES

Butenandt, A., Beckmann, R. and Hecker, E. 1961. Ueber den Sexuallockstoff des Seidenspinners. Teil I : Der biologische Test und die Isolierung des reinen Sexuallockstoffs Bombykol. *Hope Seyler's Z. Physiol. Chem. 324*, 71-83.

Kaissling, K.E. and Priesner, E. Unpublished work.

Kasang, G. 1968. Tritium-Markierung des Sexuallockstoffes Bombykol. *Z. Naturforschg. 23 b*, 1331-1335.

Schneider, D. 1970. Internat. Sympos. on Gustat. and Olfact. Geneva. These proceedings.

Steinbrecht, R.A. 1970. In Preparation.

THE INSECT ANTENNAE AS A MODEL OLFACTORY SYSTEM

Lynn M. Riddiford

Biological Laboratories, Harvard University
Cambridge, Massachusetts

The antennae of the male Chinese oak silkmoth, *Antheraea pernyi,* have specialized sensory cells (sensilla trichodea) for the reception of the female sex pheromone; these cells are not found on the female antennae (Schneider et al., 1964). The moths normally mate whenever they are placed together but did not do so for 6 hours after the male antennae were bathed in Ringer's solution for 30 minutes (Riddiford, 1970). The bathing solution contained protein, and electrophoretograms of the eluent from 20 pairs of antennae showed a sex-specific pattern with one major band found only in the male. Males were exposed to females that had been injected with H^3-sodium acetate and were emitting a volatile, lipid-soluble substance. Then after electrophoresis, radioactivity was associated only with the major male-specific protein in 2 of 8 experiments as seen in Fig. 1. This preliminary evidence thus suggests that this protein may be the sex pheromone receptor.

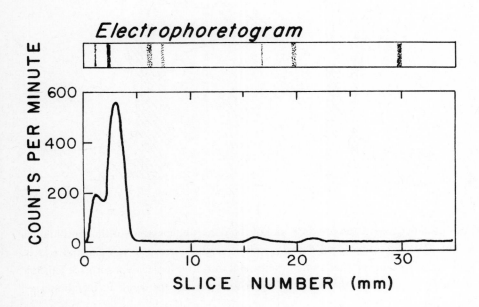

Fig. 1. Radioactivity in a polyacrylamide gel electrophoretogram of the antennal eluent from twenty male *A. pernyi* after 2 hours exposure to ten female *pernyi* moths, each of which had been injected with a total of 200 μc ^{3}H-sodium acetate. The gel was cut into 1 mm slices, ant the ^{3}H counts of each slice assessed. From Riddiford (1970).

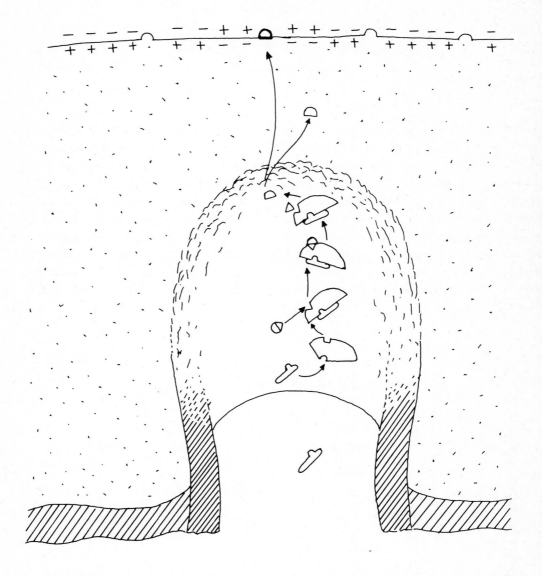

Fig. 2. Hypothetical diagram of the proximal end of the pore tubule system of the sensillum trichodeum. The odorant molecule enters, combines with the receptor protein, and acts as an allosteric effector such that a small chemical transmitter molecule is formed. This latter molecule travels across the sensillum liquor to depolarize the dendritic membrane.

Presumably, this protein comes from the pore tubule system of the insect antennae. Although the proximal extension of the tubule system is still undefined in the moths (Steinbrecht, 1969), the following model for the role of the receptor protein is based on the cytological study of the antennae of a beetle (Ernst, 1969). In this instance the proximal end of the tubule is filled with an amorphous substance and apparently terminates about 50 Å from the dendrite.

My model of the sensillum trichodeum, as seen in Fig. 2, assumes that there is only one receptor protein. The sex pheromone combines with this protein and acts as an allosteric effector (Monod et al., 1963) to trigger an enzymatic reaction which produces a small molecule. This molecule is effectively a "chemical transmitter" since it diffuses across the sensillum liquor and causes a depolarization of the dendritic membrane. This system is analogous to the peptide hormone-cyclic AMP system where the hormone causes the activation of adenyl cyclase in the target cell membrane and the subsequent formation of cyclic AMP which acts as the "second messenger" to trigger further cell reactions (Robison et al., 1968).

The sensilla basiconica found on both male and female antennae respond to many different odors by either an increased or a decreased rate of firing or by no change in this rate (Schneider et al., 1964). To account for these reactions using the above model, one has to assume a minimum of two proteins. One is similar to that in the sensillum trichodeum in that when the odorant is a proper allosteric effector, the "second messenger" that is formed causes a depolarization of the dendritic membrane, i.e. acts as an excitatory transmitter. A different odorant combines with the second protein and causes the formation of a "second messenger" which is effectively an inhibitory transmitter since it hyperpolarizes the membrane. Thus, an increased firing is due to an excess of excitatory transmitters, a decreased firing to an excess of inhibitory transmitters, and no change to either a balance of the two or an odorant which does not combine with either protein. Since a single protein may have more than one allosteric effector site, the number of receptor proteins in these generalized receptors may be small. The final integration of the three types of signals from several sensilla must be done in the brain.

REFERENCES

Ernst, K.E. 1969. Die Feinstruktur von Riechsensillen auf der Antenne des Aaskäfers Necrophorus (Coleoptera). *Z. Zellforsch. 94*, 72-102.

Monod, J., Changeux, J.P. and Jacob, F. 1963. Allosteric Proteins and Cellular Control Systems. *J. Mol. Biol. 6*, 302-329.

Riddiford, L.M. 1970. Antennal Proteins of Saturniid Moths – Their Possible Role in Olfaction. *J. Insect Physiol. 16*, 653-660.

Robison, G.A., Butcher, R.W. and Sutherland, E.W. 1968. Cyclic AMP. *Ann. Rev. Biochem. 37*, 149-174.

Schneider, D., Lacher, V. and Kaissling, K.E. 1964. Die Reaktionsweise und das Reaktionsspektrum von Riechzellen bei *Antheraea pernyi* (Lepidoptera, Saturniidae). *Z. vergl. Physiol. 48*, 632-662.

Steinbrecht, R.A. 1969. In Olfaction and Taste, ed. C. Pfaffman, Comparative Morphology of Olfactory Receptors. Rockefeller Univ. Press, pp. 3-21.

Questions from the chairman to the round table participants.

Question 1. Is it possible to consider the basic facts of taste and smell comparatively with the mammalian and insect chemoreceptor systems?

Kare :

Many illustrations of olfaction and taste have been given using the insect system as examples. Frogs eat flies, yet man does not relish them. Butterflies chemically attract one another but have no apparent chemical effect on me. Synthetic sweeteners, appealing to many humans, are offensive to many species. All the evidence supports the conclusion that species live in different sensory worlds which may or may not overlap. To rephrase the question, what purpose is there in reporting the sensory response of a frog or a moth to a chemical stimulus?

Kasang :

In my lecture I have only discussed the possibilities of enzymatic and nonenzymatic mechanisms in chemoreception of insects. Although mammals and insects have different taste and olfactory organs, the structures and functions of the chemoreceptor cells show analogies. For instance, there are no indications for differences of primary effects between mammals and insects. Primary processes in both systems seem to be physico-chemical processes without chemical changes of odor molecules.

Riddiford :

I think if one is ever going to find receptor proteins (assuming that they exist), then the insect system is probably the best one to examine. Here the cuticle makes it such that these receptor proteins must be brought to the surface in some way : either in the pore tubule system, or in the sensillum liquor, or via the tubules extending directly to the dendritic membranes. Since the cuticular proteins are insoluble, they are much easier to isolate than other types of protein. In the mammalian system, attempts have been made to isolate receptor proteins from the tongue or nasal epithelium. But many different types of protein are obtained during the isolation procedure, and then the problem of separating the ones associated with the receptors is considerable. So even though moths seem far removed from humans, perhaps the basic molecular receptor mechanism may be very similar. Then once we begin to understand the mechanism in insects, we can move up to the mammals and to man.

Cagan :

I feel that we must be very cautious in crossing species lines in these early stages when trying to assess the molecular basis for specificity. The only published data in the mammal on this point for an isolated preparation is the complexing ability of sugars with bovine tongue proteins, measured using differences in refractive index. I think any presumptive receptor molecule should be able at least to provide for specificity that correlates with behavioral results in that species, because differences exist among acceptabilities of the same series of compounds in different animals. For example, saccharin, which is generally sweet to humans, causes no behavioral response in cows and a rejection response in cats. The reported complexing data in this particular case unfortunately do not agree with the relevant behavioral data and we need to examine this aspect very closely.

Henning : For two reasons, I do not feel particularly concerned by the implications of your question. In the first place, comparative biochemistry has taught us that many metabolic reactions going on in living systems are the same everywhere, whether you take a fly or an ameba or a human being, and the same is more or less true about the functioning of various sense organs, like the eyes, and the functioning of nerve fibers. So I think it is not unreasonable to suppose that the basic mechanism is the same. Secondly, we have been talking about single chemoreceptors, but animals and human beings mostly use many receptors simultaneously, so that the recognition of the stimulus presumably depends on the recognition of the stimulation pattern across many receptors simultaneously. Since this pattern is obtained by processing of the receptor input in the neuronal network and this network can differ from one species of animal to the other, this can easily explain differences in appreciation of the same stimulus in different animal species. But I do not see why the basic receptor mechanism could not be the same in different species of animals.

Question 2. What can we realistically expect in the next decade from all the fundamental research on olfaction and taste?

Kare : One of the speakers earlier today said that perhaps in the next fifty years a further classification of compounds and their effects might be looked forward to. I am much too impatient to wait fifty years. I hope that the seventies will give us an understanding in this area. I am reasonably certain it will so far as the chemoreceptors are concerned, but I would like to ask the other member of the panel for their comments on this question.

Henning : When we look at the history of the work which has been done on olfaction and taste, we see that it started with psychophysics, and I feel very sorry to say that in the seventy years or so of its existence, psychophysics has not taught us much. You can easily see this in the attempts to correlate structure with odor or taste, and if there is any conclusion we can draw from all this work, it is that there is no simple relation between molecular structure and odor and taste. Electrophysiology has made very important contributions and will continue to do so, both regarding receptor processes and the coding mechanism, provided we find methods for recording from many cells simultaneously. But in my opinion no real advance will be made if we do not develop proper biochemical techniques and do not integrate biochemical knowledge into the field. This seems to me a most important point, and the way we have to go in the next decade.

Cagan : I am sorry to also disappoint you about the timing of what we are going to achieve in the next few years. Perhaps I can elaborate by continuing to use an analogy with two related fields, the historical perspective of which we should bear in mind. Enzymes have been studied intensively since the 1920s and it was in 1926 that Summer crystallized the first enzyme; purification of intracellular enzymes did not begin until the late 1930s, but since then, literally hundreds of individuals, and in fact entire research institutes have studied enzymes exclusively and intensively. Although much is known, many of the subtleties still escape explanation. More recently, membrane transport proteins have been studied. Perhaps we can find some encouragement in the fact that it was only in 1965 that the presence of a membrane transport protein was first demonstrated. In the last few years, the presence of several other such proteins has been demonstrated, but the intensively sought after acetylcholine receptor still eludes us. Hopefully we will be able to draw upon the experience in these other fields to advance our own, we might expect progress at the molecular level in explaining specificity, but we should not expect, in my opinion, that simply isolating one type of receptor molecule will answer all our questions.

Riddiford :   I might add that even when we do get a receptor protein or a receptor molecule from whatever source, we are going to have some problems working with it, because it seems as if, at least in both my and Dr. Kaissling's preliminary experiments, no protein eluted from the antenna can bind the tritiated pheromone *in vitro*. This is not unreasonable, because once out of their particular environment, proteins change their conformation. They may be aggregated *in situ,* or they may be membrane-bound. Then once removed from the membrane, their conformation changes so the particular receptor site is no longer exposed. So the techniques of membrane protein chemistry which are just becoming known are going to be very important for examining the types of receptor molecules we are discussing.

Kasang :   We cannot assume that in the next decade all processes in olfaction and taste can be explained. These problems are too complex as we have seen. Going on from insects, by electron microscopy and autoradiography some details of membranes of receptor cells could be analyzed. With specific odor substances and inhibitors, by electrophysiological and biochemical methods we will get more information about structures and functions of receptor membranes. By new physico-chemical methods like fast reaction recording during interactions of odors with acceptors we will obtain more knowledge about the important primary processes.

Question 3. Of what practical consequence is all the research work we are doing?

Kare :   I am optimistic that this decade will be the golden years for the chemical senses for a number of reasons. Our major biological problem now is the population explosion and with it, problems of hunger, which are very closely related to taste and smell. We can incidentally contribute to the quality of life while we are looking at the fundamental problems.

I think that it might be worth asking one last question. Our host, Firmenich, as well as the other companies in this area, are interested in the fundamental aspects of our research. Might I ask the participants of what practical consequence is all the work we are doing?

Cagan :

One good question deserves another and I can answer partially by posing a question. How far and how rapidly can chemists continue to design compounds with particular flavors and fragrances, knowing nothing about the receptors with which they interact? Pharmaceutical chemists have asked this question for some time and have been actively engaged in work designed to uncover the nature both of the receptor sites and also to design drugs to interact specifically with them, and I think this is one important reason for continuing along this line. Another intriguing possibility is that of analysis or assaying such a physiological effect, but in an *in vitro* system. I am reluctant to predict that we shall soon have such a simple biochemical assay for a particular flavor, but this is one direction that increased fundamental knowledge of receptors itself could possibly lead.

Kasang :

Our work with highly active and specific insect sex attractants leads us to think of practical applications in biological pest control. For the measurement of small amounts of odorous substances new analytical methods were developed. Probably the most important practical applications in future will be in medicine, psychology, and sociology.

Riddiford :

The people working with foods and perfumes, probably wonder about the relevance of working with insects and insect olfaction. There is one practical consequence that has not been mentioned. I think that the pheromones are going to be increasingly used as a means of insect control. There are possibilities of using masking odors to block perception by males of the female sex pheromone and so preventing reproduction. There have already been methods of using the isolated female sex pheromone to attract the males after which the males are sterilized and released. More and more vertebrate pheromones, particularly mammalian are becoming known. Even though we are not aware of them, they may influence our life in many subtle ways. It is not certain that the nose plays a role in human social interactions; but, if so, air pollution might present problems other than a serious health hazard.

Henning :

Dr. Kare, I thought your view was optimistic; mine is just as optimistic, although I do not know on what time scale I am speaking, years, or tens of years, but let us hope the former. A good many of the practical problems we face in perfumery or in food flavoring presumably have their origin at the receptor level. We can ask ourselves why saccharin is 400 times sweeter than sugar or how flavor potentiators work, we can try and suppress certain tastes we do not want, and we can ask ourselves whether it is necessary to incorporate the 300 compounds you can find in coffee flavor into a mixture to imitate this flavor, or whether we can do with only a few? We can also aks what are the molecular characteristics that determine the odor or taste? I expect that many of these problems will find their solution once we have a good insight into the receptor process. It will then be possible to synthesize a new molecule with an odor we want, and we will then even be able to control the flavor of our products automatically. I think that fundamental research in this field is necessary from the practical point of view. Progress will be slow, but this will be more than compensated for by the insight we will have gained and which we can apply.

Fischer :   Although you pay lip-service to progress, Dr. Kare, you should not forget that at the height of his career, Rutherford discounted the possibility of atom-smashing and the release of atomic energy. Even experts have no window to the future.

In spite of the inaccuracy of Erlich's lock-and-key hypothesis to account for the specific affinity of dyes and drugs to receptors (*"corpora non agunt nisi fixata"*), the idea initiated an era of provocative discoveries when thousands of effective drugs were discovered. Evidently both "binding sites" and "active sites" have to be considered because of "the versatility of the binding sites on proteins towards structurally unrelated ligands" (Glaser, 1970). These drugs have not, of course, specific but multiple activities, the desired activity being called the "main effect", the others "side effects". Apart from this, what is called the "mechanism of action" is largely unknown, even of such simple drugs as aspirin, swallowed by the ton every day.

Lastly, let me make some unfriendly remarks about the term "psycho-physical correlation". The use of this outdated concept is, I believe, scientifically unjustifiable, because "psyche" is undefined, but in spite of this it is correlated with something measurable.

Kafka :   Dr. Riddiford, you reported an odorant-catalyzed reaction forming a "second messenger". Where is specificity finally determined? Is it at cell level, or do you think specificity is determined within the tubules or the liquor, meaning that all the cells bathed in the same liquor of a single sensilla trichodea should have the same odor spectrum.

Riddiford :   The specificity lies with the protein having the allosteric effector site that causes the second messenger, so the first signal here is the specific one. I suppose that all the sensilla trichodea (but not the sensilla basiconica) would have the same proteins.

Kafka :   So it is still an open question whether your protein is coming from the tubuli or the sensillum liquor; it could also come from the cell level.

Riddiford :   Yes, it could come from the sensillum liquor, or it could come (although I do not think it does, unless it is weakly bound) from the dendritic membrane.

Zottermann :   I am very optimistic about the next decade because the new instruments we have will give us so much new data. In a few years we shall be able to record from single cells. When we can examine sensory organs directly with the microscope, it would be extraordinary if we did not find out what was going on.

Ruzicka :   I believe that to expect any result of value in a very short time is exaggerated optimism. Although there is only one fundamental chemistry, the chemistry in the laboratory following the same laws as the chemistry in living systems, the two utilize completely different methods. In the laboratory we use acids, but life uses enzymes. We still do not understand a single enzyme reaction, and we need a long, long time with enzyme chemistry before we can deal with the sort of question that has been discussed here.

*Numbers followed by an asterisk refer to pages on which the complete reference is listed.

* Numbers followed by an asterisk refer to pages on which the
  complete reference is listed.

Dravnieks, A., 42, 125, 145, 177
Duncan, C., 220, 228*
Duncan, R.B., 13, 22*, 239, 242*
Dzendolet, E., 74, 85*, 222, 228*

E

Easter, W., 167, 174*
Easty, G., 199, 227*
Eccles, J.C., 90, 90*
Eckert, L., 200, 228*
Egli, R.H., 94, 108*, 173, 175*
Eisner, T., 53, 54, 58*, 59*
Ekman, G., 87, 90*
Elsberg, C.A., 88, 90*
Emslie, A.G., 134, 143*
England, S., 213, 230*
Erdtman, H., 169, 174*
Erickson, R., 221, 228*
Erlenmeyer, H., 208, 216, 228*
Erman, W.F., 167, 175*
Ernst, K.-D., 47, 58*, 61, 69*,
     253, 253*
Eschenmoser, A., 180, 183*
Eugster, C.H., 172, 175*
Evans, H.G.V., 13, 25*
Eykelboom, A.J., 43, 236, 259
Eysenck, H., 207, 208, 228*
Eysenck, S., 207, 208, 228*

F

Fagerson, I.S., 99, 104, 105*
Ferguson, J., 198, 228*
Fieldner, A.C., 140, 143*
Filshie, B., 199, 228*
Fink, R.F., 15, 19, 21, 24*,
     25, 26*
Fischer, R., 43, 80, 85*, 86,
     187, 223, 224, 228*,
     229*, 230*, 231*, 232*,
     235*, 258, 259
Fisher, B.E., 172, 175*
Fisher, R., 188, 213, 223, 231*

Flament, I., 173, 175*
Forss, D.A., 104, 105*
Fox, A., 212, 231*
Franzén, O., 90*
Freeman, J.A., 240, 242*
Freud, S., 187, 231*
Friedel, P., 173, 174*
Frisch, D., 29, 37*
Fromms, S.P., 6, 9, 24*
Fujimaki, M., 94, 105*, 109*
Furia, T.E., 107*

G

Gale, A., 207, 231*
Galton, F., 225, 231*
Gander, J., 202, 231*
Gardner, W.H., 100, 101, 105*
Garn, S., 213, 230*
Garten, S., 2, 22*
Gasser, H.S., 29, 37*
Gautschi, F., 173, 175*
Gebbie, H.A., 141, 143*
Gentili, B., 99, 104, 106*
Gerebtzoff, M.A., 2, 23*
Gesteland, R.C., 20, 23*, 30, 36,
     37*, 134, 137, 143*
Getchell, T.V., 30, 37*
Gianturco, M., 92, 105*, 173, 174*
Giersch, W., 178, 180, 183*, 184,
     186*
Gilman, A., 208, 231*
Girgis, M., 226, 231*
Glanville, E., 204, 205, 217, 232*
Goedde, H., 216, 231*
Gold, H.J., 171, 175*
Goldkamp, A.H., 94, 99, 107*, 200,
     233*
Goldman, I.M., 173, 175*
Good, R., 172, 175*
Goodman, L., 208, 231*
Gordon, H., 221, 231*
Granit, R., 90, 90*

*Numbers followed by an asterisk refer to pages on which the
   complete reference is listed.

* Numbers followed by an asterisk refer to pages on which the
  complete reference is listed.

* Numbers followed by an asterisk refer to pages on which the complete reference is listed.

* Numbers followed by an asterisk refer to pages on which the
  complete reference is listed.

* Numbers followed by an asterisk refer to pages on which the
  complete reference is listed.

R

S

* Numbers followed by an asterisk refer to pages on which the
  complete reference is listed.

* Numbers followed by an asterisk refer to pages on which the complete reference is listed.

* Numbers followed by an asterisk refer to pages on which the
  complete reference is listed.

SUBJECT INDEX

A

Age, influence on taste threshold, 205

Absorption spectra and olfactory responses, 135

Aglycogeusia, 131

AH,B system, 128

Albinos, 12

Alpha-helix, see ⍺-Helix

Almond odor, 147

Ambergris odor, 178

Amygdala, electrophysiological activity of, 39, 226

Anosmia, 2, 153

Antheraea pernyi, 251

Aphrodisiac, 53

B

Bee, pheromones of, 46

Benzaldehydes, odor of, 155

Binding mechanisms, 67, 128

Bitter taste, 129

Bombykol, 46, 245
    metabolites, 246
    tritiated, 48, 245
    metabolic transformation on antennae, 246

Bombyx mori, sexual attractants of, 46, 245

Bowman's glands, 6, 13

Brain potentials after stimulation of tongue, 81

Bulbar unit activity, 20

Butterflies, pheromones of, 52

C

β-carotene, 13

Carrots, flavor of, 97

Cheese, flavor of, 98

Chemical
    communication, 45
    signal compounds, see Pheromones

Chemoreceptors of butterfly, 52

Chemoreception,
    genetic aspects of, 212
    models of, 198
    and the olfaction/gustation interaction, 221

Chromoproteins, non-carotenoid, 14

Citronellal, 116

Coffee odor, stimulation by, 88

Crassius crassius, 10

Crastia amymone, 53

D

Dacus dorsalis, 135

Danaidae, 52

Danaus gilippus, 53

Danaus plexippus, 55

Decalin ring compounds, 178

Direct taste effects, 92

E

EAG, see Electroantennogram

EEG, see Electroencephalogram

Electroantennogram, 53

Electroencephalogram, "averaging" of, 73

Electro-gustometer, Figs 4-6, 77

Electro-olfactogram, 30, 88

EOG, see Electro-olfactogram

F

Fatty acids, odor of, 149

Flavor spectrum, 92

Florida Queen-butterfly, see Danaus gilippus

Food flavors, composition of, 92

Frog, 8, 20, 34

Fruit, sugar-acid ratios, 100

Furanones, 172

G

Guitar fish, see Rhinobatus lentiginous

CORRIGENDA

P. 24, ref. 9, for "Pfaffman", read "Pfaffmann".
  ref. 14, for "Shibuya T.S.", read "Shibuya T.".
p. 29, line 2, for "Helst", read "Heist".
  line 31, for "Casser", read "Gasser".
p. 36, line 3 from bottom, for "Oethier", read "Dethier".
p. 38, ref. 13, for "Tucker S.", read "Tucker D.".
p. 85, ref. 7, for "Jastrau",read "Jastram".
p. 104, para. 2, last line, for "Mackey", read "Mackay".
p. 106, refs. 14, 15 & 16, for "Kininaka", read "Kuninaka".
p. 108, ref. 12, for "Takamoto", read "Takemoto".
p. 121, line 2, for "Appell", read "Appel".
p. 144, ref. 4, for "Sturart", read "Stuart".
p. 176, ref. 1, for "D.F. Seidel", read "C.F. Seidel".
p. 184 and p. 186, ref. 1, for "McLeod", read "MacLeod".
p. 187, last line, for "Kaebling", read "Kaelbling".
p. 212, line 10, for "1949a", read "1949".
p. 219, para. 4, line 3, insert the ref. "Schönpflug (1969)".